Boris Menin

Drum Freezers: Computer simulation and application

AF138641

Boris Menin

Drum Freezers: Computer simulation and application

Rational design and manufacturing of drum freezers

LAP LAMBERT Academic Publishing

Impressum / Imprint

Bibliografische Information der Deutschen Nationalbibliothek: Die Deutsche Nationalbibliothek verzeichnet diese Publikation in der Deutschen Nationalbibliografie; detaillierte bibliografische Daten sind im Internet über http://dnb.d-nb.de abrufbar.
Alle in diesem Buch genannten Marken und Produktnamen unterliegen warenzeichen-, marken- oder patentrechtlichem Schutz bzw. sind Warenzeichen oder eingetragene Warenzeichen der jeweiligen Inhaber. Die Wiedergabe von Marken, Produktnamen, Gebrauchsnamen, Handelsnamen, Warenbezeichnungen u.s.w. in diesem Werk berechtigt auch ohne besondere Kennzeichnung nicht zu der Annahme, dass solche Namen im Sinne der Warenzeichen- und Markenschutzgesetzgebung als frei zu betrachten wären und daher von jedermann benutzt werden dürften.

Bibliographic information published by the Deutsche Nationalbibliothek: The Deutsche Nationalbibliothek lists this publication in the Deutsche Nationalbibliografie; detailed bibliographic data are available in the Internet at http://dnb.d-nb.de.
Any brand names and product names mentioned in this book are subject to trademark, brand or patent protection and are trademarks or registered trademarks of their respective holders. The use of brand names, product names, common names, trade names, product descriptions etc. even without a particular marking in this work is in no way to be construed to mean that such names may be regarded as unrestricted in respect of trademark and brand protection legislation and could thus be used by anyone.

Coverbild / Cover image: www.ingimage.com

Verlag / Publisher:
LAP LAMBERT Academic Publishing
ist ein Imprint der / is a trademark of
OmniScriptum GmbH & Co. KG
Heinrich-Böcking-Str. 6-8, 66121 Saarbrücken, Deutschland / Germany
Email: info@lap-publishing.com

Herstellung: siehe letzte Seite /
Printed at: see last page
ISBN: 978-3-659-66917-0

Zugl. / Approved by: Beer-Sheba, 2014

In memoriam Guygo E.I and Fomin N.V.

Drum Freezers:

Computer simulation and application

Dr. B. M Menin

In the book there are systematically choosen materials necessary for rational design, construction and operation of drum freezers that are used in various branches of engineering. The theoretical basis of heat and mass transfer in the process of freezing of the material on the cooling surface is discussed. Various designs of apparatus of a continuous drum type and the materials used in their manufacture are discribed.

The book is intended for engineers and technicians of project offices and companies involved in the design and using of refrigeration equipment.

2014

CONTENTS

Basic symbols

a – temperature conductivity, m^2/s;

c '- specific heat capacity, $J/(kg \cdot K)$;

c - specific volumetric heat capacity, $J/(m^3 \cdot K)$;

D - outer diameter of the drum evaporator, m;

F – surface, m^2

h - the thickness of the ice layer or frozen material, m;

l - linear dimension m;

n - frequency of rotation of the drum evaporator, s^{-1};

r - the latent heat of phase transition, J/kg;

r_v - specific volumetric heat of the phase change, J/m^3;

R - the wall thickness of the drum evaporator, m;

t - temeratura, °C;

v - velocity, m/s;

w – amount of frozen water at a given temperature,%;

W – product humidity,%;

x, y - coordinates, m;

α - heat transfer coefficient, $W/(m^2 \cdot K)$;

λ - thermal conductivity, $W/(m \cdot K)$;

v - kinematic viscosity, m^2/s;

ρ - density, kg/m;

τ - time, s;

ξ - extent of the zone of freezing and hypothermia of material, m

FOREWORD

Drum freezers are widely and successfully used in the fish, meat and dairy, food processing and other industries, in practice, in most countries of the world. Over the past few years, the considerable efforts are made to improve their design and effectiveness. However, information on the results of these studies can be found only in scattered publications and periodicals that impede to generalize and use the data obtained in the practical engineering.

The basis of the book is based on a real, experimental and theoretical investigations carried out by the author for 35 years, as well as development experience and extensive industrial use of these devices on board fishing vessels and shore facilities accumulated by users over the world.

The book presents methods for engineering design of drum freezers of continuous action on the basis of mathematical models of heat transfer processes and the development of computing experimental systems that meet the capabilities of modern computing engineering.

The limited size of the book does not allow an exhaustive review of all matters of the design of drum machines; deliberately omitted and such is certainly important issues such as choice of refrigeration systems schemes and their automation, especially as these issues in sufficient detail described in a number of monographs.

The author set himself a slightly different problem: based on analysis of existing constructions of drum freezers to substantiate the feasibility and necessity of development of theoretical models , the practical significance of methods of the general mathematical description of the processes of heat transfer in the machines for the their focused improvement.

The author hopes that the book may be useful for engineers working in the enterprises and organizations of the corresponding profile and students of universities and colleges.

Comments and suggestions on the content of the book should be sent to the following E-mail: meninbm@gmail.com

INTRODUCTION

In today's environment, the need of the prolonged cold storage and transportation of large quantities of food has been one of the most important priorities. The scale of the problem makes it necessary to create technologies and means of high performance, low power consumption and material consumption for cooling and freezing of foods.

Even in ancient times people knew in their practice that the effect of low temperatures prevents spoilage, can store them for a long time.

List of food products and other materials to be processed by refrigeration, particularly, freezing is currently very wide. Processing objects are different in physical-chemical properties and structure (solid, liquid, paste and minced-shaped, homogeneous and heterogeneous), dose size (small-size or having a significant size). Accordingly, different types of freezers are used for this purpose. According to various estimates, there are freezers of not less than one hundred types in the world.

To some extent this is understandable. At the present stage of the technological development, it is not possible to construct a universal freezer fully satisfying the requirements that may be presented for its construction and operation modes for all applications. However, it is obvious that the number of types of devices can be substantially reduced by their unification on the basis of a scientific approach to this approach.

The lack of a scientific theory, which determines the validity of the choice of optimal values of technical and economic parameters of freezers, puts the solution of this problem in reliance on intuition and experience of the designers of devices. During the design of the machines do not always take into account the substantial connection between the requirements concerning their use, and the conditions of exploatation; often not considered a sufficient number of different design and technology options. Thus, the question of optimal design decisions and the principle of the effectiveness of design alternatives is still open. Therefore, at present there is strong tendency to turn to the system principles of problem solving in the field of design.

From the point of view of system analysis, Freezer is a technical system for cooling, freezing and hypothermia of various substances. The purpose of the system of research is to create a model of the studied object, which at the same time considering all its components, their properties and relationships.

Systematic study of a complex object defines a fundamentally new direction in the development of the methodology of modern scientific research. If before it was on the description of the object, and the knowledge was designed to study some of its properties, using a systematic approach reveals the mechanism of operation of the object, taking into account its internal and external characteristics.

Despite this advantage to the system approach and the large number of publications on the formalization of construction machines and apparatus, this problem is not completely solved. For example, proper selection of the system object of its hierarchical structure (in particular, the selection method of heat treatment material subjected to refrigeration, the choice of configuration of the working volume occupied by these pictures, etc.) is usually carried out on an intuitive basis.

The urgent need to replace obsolete by new designs with justification feasibility of specifically chosen method of cold treatment of the material causes the need to study the so-called "life-curves" of technical systems and their laws of development [1].

Knowledge of the characteristics of "life curves" (change in time of the main characteristics of freezers - productivity, the number of production units of this design, energy consumption, etc.) is required to select one of two alternatives: the study and improvement of known structure or the creation of a fundamentally new device for perform a given operation.

Different technical systems have specific areas that are common to all systems and shown schematically in Fig. 1.

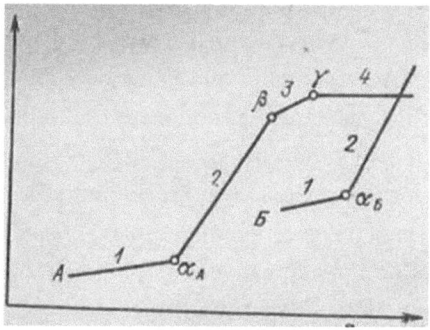

Fig.1 "The life curve" of the technical system

Since the introduction (section 1), the system evolves slowly but steadily until α_A, i.e. until the transition to mass production. At the point of β, the possibilities of the used principle in the device are exhausted, and found it inexpedient to further economic development: on phase 3, its development growth begins to subside. After a point γ system A is degraded or replaced by a fundamentally different system B, or for a longtime keeps achieved results.

Creation of a freezer design is due primarily to the needs of its economy, the level of technology, but also the amount of funds invested in research and development.

The latter, apparently, explains the long time interval during which the creation of new designs of devices: from J. Lespi built in 1810, "icemaker" that worked on the principle of a steam vacuum machines and up to currently designed icemaker-clarifier (desalinator) of the natural rain waters. A significant obstacle to the unification of freezers is that their applications overlap only partially due to the large variety of materials subjected to cold treatment.

Air freezers as a technical system, it seems, passed the point β. Still widely their usage is explained because of freezing of products by air allows to keep their satisfactory nutritional properties and appearance. As a rule, one and the same apparatus can be used for cooling and freezing of food of various geometric shapes and sizes. Significant disadvantages of these devices should include a relatively low coefficient of heat transfer from the product to the air - about 50 $W/(м^2 \cdot K)$, the uneven freezing of products , a small rate of freezing of the product - about (0.65-1) cm/h, high specific indicators of weight - up to 1.56 t/(t/d) , energy consumption - up to 115-200 kWh/t (pure ice), product shrinkage - (1-2) %.

The usage of plate freezers (horizontal and vertical type) was dictated by the desire to eliminate the intermediate environment. It helped to reduce the temperature difference and the enhancement of heat transfer between the refrigerant and frozen products (coefficient of heat transfer from the product to the refrigerant was about 300 $W/(м^2 \cdot K)$, the possibility of refusing bulky and heavy air- coolers and power-hungry fans used in air devices. According to our opinion, plate freezers as a technical system are located in the area α - β of "life curve".

The main drawback of most plate freezers is the periodicity of action, which limits their performance. To eliminate this drawback, various solutions have been proposed, including the design of a rotary freezing plate apparatus. However, these freezers fell to expectations. In particular, the use of rotary units instead of air on board of the fishing fleet has resulted in poor performance of freezers by approximately 20% and an increase in the number of staff by 1.5 times. This situation can be explained by the fact that the invention, in its design aimed for improving the system, by the implementation of the idea at the incorrect technical level, is not always conducive to the improvement of this system.

9

A new design of the "pendulum type" device will help to expand the scope of plate freezers. It combines the advantages of vertical and horizontal plate freezers: loading and unloading of product is in the horizontal position and the actual process of freezing is in a vertical position. This creates a good basis for a significant intensification of heat removal from the frozen food, as well as the mechanization and automation of the product loading and unloading.

At the initial stage of development there is a technical system as contact freezers, in which the direct intense heat is removed from the frozen food to the heat medium (liquid nitrogen, air, carbon dioxide, Freon-12, passed a special chemical treatment, an aqueous solution of sodium chloride and etc.). Unfortunately, in this principle there are incorporated features that impede the spread of such devices and, ultimately, which can cause its rapid degradation and "aging". These include significant expenditure of the non-reproducible heat medium, its high cost, the deterioration of the treated products and, in part (for using CFCs), and damage to the environment.

The family of freezers constructions is certainly not limited to the three mentioned (with numerous modifications) groups. However, if we restrict ourselves only to those groups, in each of them there are contradictions, the solving of which will influence on the prospect of the development of the system or the feasibility of transition to a different system, which includes the former as a subsystem.

In the works to improve the old and to develope new designs of devices, many groups of scientists, engineers, and designers are involved. How to organize their creative exploration, as will be chosen techniques to achieve the formulated targets? It determines the quality of the design of refrigerating equipment, the savings in the use of research resources. Requirements applicable to the apparatus are largely inconsistent. The designer should consider, so to speak, thousands of options. This leads to the fact that often selected is not the best solution.

One of the areas of overcoming these difficulties are to use the theory of inventive problem solving, which was founded by the Soviet engineer G.S. Altshuller. According to this theory, the development of any technical system is subject to certain rules that must be fulfilled in order the system is viable.

Thus, the law of completeness of technical system is expressed in the fact that a necessary condition for the fundamental viability of the system is the availability and performance of its main parts. For freezers, this means that any unit in the operating position must include a proper freezable product, refrigerant, transmission (transmission's body), and a control unit by which the changing of the state of the products is regulated. Moreover, the controllability of the system can be ensured only if at least one part of it will be managed.

These, at first glance, obvious recommendations are not always taken into account. For example, under technology developed in the 1960s in the U.S., foods prepared for freezing are fed by conveyor into the vessel with liquid nitrogen. Then frozen foods supplied by the conveyor directly to the packaging location. The vaporized nitrogen gas is liquefied and returned back into the machine.

Currently, these devices are not used, because due to too rapid freezing of the product it is difficult to adjust its length of exposure in the bath, and hence evaporation of nitrogen. Such devices are extremely wasteful in comparison to devices that implement the method of irrigation products with liquid nitrogen flow rate of which can be easily adjusted. The time interval between the filing of patents for these freezers (5-6 years) is the price paid for not knowing the laws of technical systems.

A prerequisite of basic viability of the system is the pass-through of energy in all its parts (the so-called "law of energy conductivity of the system"). Transmission (transfer) of energy from one side of the machine to another can be material (contact freezing of the product in the refrigerant), field (by conduction in the tile devices) and material-field (by taking away the convective heat flow of air moving in air freezing apparatus). From this standpoint, plate devices have undoubted advantage in comparison with air freezers where there are substantial energy losses due to heat of the cooling air, i.e. not all absorbed energy is used to produce the desired effect: freezing of the product to the desired final temperature. When comparing the plate and contact devices for the same reason, preference should be given to the latter.

One of the objective laws of the development of technical systems (including freezers) is that systems with mismatched rhythms are replaced by more sophisticated systems with a coherent rhythm. Let's consider the rotary freezer. This system includes the individual elements like freezer sections, loading and unloading devices, hydraulic pump and the circulation circuit, etc.

Pump-circulating circuit is designed to provide a constant flow rate of refrigerant in any of the freezing plates. However, the heat load from the products within each cycle is unstable and decreased from the maximum value that is created at the loading of the each next portion of the product to a minimum while unloading frozen products. Thus, at the moment of peak the pump-circulation scheme does not provide "the required cooling capacity" in the freezing plates, and at the end of each cycle, only a small part of the "cold potential stored by refrigerant" is efficiently used: part of the energy is spent on "spurious" heat flow.

Listed here are just some of the laws that determine the development of the technical system. Using the "code" of such laws, it is possible to formulate the requirements for the design of an ideal freezer unit. Obviously, the ideal can be considered as a device that provides the specified performance and operational parameters of the process with minimum weight, volume and power consumption.

In practice, there is a paradox: the developed new freezers are becoming more and more large-size, energy-intensive and heavy. This paradox is explained by the fact that the natural tendency of increasing performance of the machine is often not accompanied by improving how it works.

Difficulties in the development of requirements for the ideal machine are also exposed to a variety of refrigerated materials (their shape, size, mechanical, thermal and moisture, weight and metabolic characteristics).

Based on the foregoing, we can formulate the requirements that should guide the designer when creating a freezer unit. These include the continuity of the technological process of cold treatment of materials; the existence of variable "energy connection" between the heat load and flow of the consumed cold; changes in the structure or properties of the product to be treated in order to reduce the cooling load; the application of endothermic processes, providing a high intensity of cold treatment; reduction of the volume occupied by the unit accompanied by increasing a specific capacity of apparatus.

In many ways, the prototype of the ideal freezer unit can serve as a drum freezer. It represents a technical system, in which the above requirements are most fully realized.

Accordingly, drum machines in comparison with other types of freezers are characterized by the lower specific energy consumption (0.056 kWh/kg), the lowest required cooling and freezing energy (less 0.139 kWh/kg) and the lowest specific volume occupied by unit (0.3 m³/(t/day)).

The drum freezers provide a high preservation of the initial properties of the material due to the uniformity of the temperature field through its thickness; they can work in automatic continous mode that reduces labor costs. Thanks to these advantages drum freezers are becoming more widespread.

CHAPTER 1. APPLICATIONS AND DESIGN OF DRUM FREEZERS

1.1. Purpose and scope

In the list of refrigeration technological equipment used in industry, drum freezers (DF) most widely known as flake ice generators and apparatus for freezing food pastes, minced, small-piece products and other water-containing materials for various purposes. In recent years there have also drum machines are specially designed not used for freezing wet materials and products, but for temporarily fixing them on the surface of the drum in the implementation of a number of technological processes. Designed apparatus are for thin-layer removing the skins from the fish fillets (skinning machines), for keeping the non-magnetic components in processing, such as grinding, etc. We can expect the use of such devices and method of keeping materials on the freezing surface in sorting devices, various feeders, and robotics.

However, the main purpose of DF, discussed below, is the freezing or cooling materials in a continuous stream. Continuity of the process, the intensity of heat transfer in thin layers, simplicity of the device and the service light "inlining" in manufacturing lines contribute to the fact that these devices are widely used in different industries.

The greatest number of DF for the production of ice is used in the fish, meat and dairy, food and chemical industries, as well as in construction, medicine and instrumentation. All becomes wider use of DF in the process of cryo-concentration of aqueous solutions, for salt water desalination, purification of drinking water and wastewater.

1.2. Basic design

Regardless of the diversity of types of DF, they all have one common and main element - cylindrical freezing evaporator. It can be movable or stationary, horizontal or vertical, and with different diameter and length.

The use of the contact method of rapid freezing on the drum surface can dramatically reduce the loss of product shrinkage. During the test of the fish fillet, it was only 0.2% instead of 1% for the best air continuous action freezers.

Horizontal DF are designed and manufactured for use in the chemical industry in the manufacture of latex (this is the raw material for the rubber industry) and for the treatment of natural sediments and sewage. The design of these devices is based, essentially, on the same principles: usage of brine or another antifreeze liquid for the drum cooling and removing of the frozen product by the scraper.

Thus, in practice, at the development of DF for freezing pastor and minced meat products there is a trend of the revival of systems with horizontal drum and brine cooling.

However, in the design of developments during recent years there are taken steps to intensify the processes of heat transfer, in particular, due to the transition to the scheme of the directly boiling refrigerant in a narrow channel or by increasing the flow rate of coolant in the slot gap.

Currently, the number of countries and manufacturers of DF, primarily in the form of flake ice generators, grows. Analysis of patent, technical, financial, and R&D documentation of the development of this refrigeration industry over a sufficiently long period, shows that in the next 10-15 years, there are not expected other solutions in the field of thin-film continuous freezing of liquid and small-piece products. It can be assumed that the main type of devices for these purposes will remain freezing drum apparatus. However, it is necessary to conduct further searches for ways to intensify and improve the technical and economic performance of these devices. For example, do not stop work on the study the impact of physical fields (magnetic, electric) on the intensity of heat transfer processes. In principle, it is possible to use drum machines in the range of cryogenic temperatures for freezing of water-containing products, or to solve the problems of cryogenic technology, for instance, to remove ("freezing") carbon dioxide from the air.

Expanding the use of DF requires the creation and improvement of the theory of calculation and mathematical analysis of freezing process for liquids, paste, minced-shaped and small-piece products by these devices. The following chapters present modern theoretical concepts of heat transfer processes in DF and give examples of the use of theory in engineering calculations and research.

CHAPTER 2. HEAT TRANSFER AT THIN LAYER FREEZING OF MATERIAL ON THE COOLING SURFACE

2.1. Physical and mathematical basis for describing process of material freezing

The process of transition of liquid to solid is a whole complex of phenomena (thermodynamic, thermal, diffusion, hydrodynamic, etc.). We can note two main directions of researching.

The first direction is considering a transition from liquid to solid state from the standpoint of the formation of crystals of solid atoms and molecules of the liquid, and finds out the cause of the dependence of crystal growth in different directions and laws governing the construction of crystal lattices and the like, the second considers the problem only with the inclusion of thermal interactions between the solid and liquid phases and the bodies which are in contact with them. With the first direction there is connected the concept of "crystallization", with the second - "hardening".

Thus, the crystallization is understood as a creation of conglomerate of crystallites at the transition of substance from the less stable thermodynamically to more stable state. Crystallization of liquids is a fixation of some arrangement of the particles relative to each other when, at a sharp reduction in overall energy of their movement, they are mutually oriented in a certain order, inherent in their position in the crystal. This substance acquires other atomic and molecular structure and new physical properties but without pronounced chemical changes [2].

Solidification is transition of matter from the liquid to the solid state, conventionally considered without considering the peculiarities of formation of crystals. Although the term "solidofication" is not a very common and well-established, there are many papers of kinetics of crystallization and solidification. In [3], it is shown that the determining factor in the process of transformation of matter from liquid to solid is the solidification process, because the transformation duration depends on heat transfer. Crystallization adapts to the heat transfer by supercooling, causing one or another structure of the solid phase.

When removing heat from the liquid its temperature and the energy of motion of the particles decrease. At a fixed energy of thermal motion, the interaction orientation of the particles begins. Crystallization centers are formation of a certain amount of the stable connected and correctly oriented with each other, particles. On these centers, the further orientation of the surrounding fluid particles and crystal growth are realized.

Thus, the crystals can occur from finished nuclei in the supercooled melt or supersaturated solution of the substance. They may be either a crystal particle of the substance, or with another substance of similar structure. It is also possible that colloidal particles could be nuclei that are capable to adsorb the crystallizing substance. The presence of various irregularities (steps, scratches, and outputs the dislocation, etc.) on the surface, which is contacted with the supercooled liquid, favor for nucleation. This is also facilitated by the presence of the mass movement of fluid, temperature fluctuations, the concentration, the properties of the boundary phase and etc.

The liquid can be cooled to a temperature below the cryoscopic temperature without formation os crystals at a small value of orienting forces. However, the crystallization can begin only at the actual supercooling (supersaturation) of melt (solution), not less than a certain threshold value. Large germ particles require for their growth smaller and small particles higher supercooling [4].

Depending on the process conditions of subcooling (purity fluid immobility, etc.), the overcooling values, for example, the water crystallization may range from $(0.1-0.2)^0C$ to 40^0C. The factual lack of subcooling is due to the influence of insoluble particles and surface-active impurities.

Further discussion will be devoted to coverage of various aspects of the solidification process materials in drum freezers. Therefore, although the crystallization of liquids in food has the features set out in numerous papers [5, 6, 7], their consideration in this book falls.

Research of any physical phenomena or processes, including solidification process of materials begins with the establishment of the simplest experimental facts. By them it is possible to formulate the laws governing the investigational material object, and write them in the form of some mathematical relationships. Volume of prior knowledge, goals of analysis, the expected completeness and accuracy of the required solutions determine the degree of schematization of process studied.

In general, the model of the solidification process does not have to contain the wording of causal relationships between the elements of the studied object in the form of finished analytic expressions. In some cases we have to be satisfied with such connections (qualitatively and quantitatively) that characterize the material object only in the most general terms, and express a significantly smaller amount of knowledge about the internal structure of the studied process. In all cases, the model of the solidification process represents intuitively selectable abstraction primarily because it is built for the intuitive assigned object, as well as due to the incomplete or inaccurate knowledge (conscious simplifying) about the laws of nature.

In most papers devoted to solving problems of heat transfer during solidification of materials, based on the Fourier law of a linear relation between the heat flux q and the temperature gradient is used:

$$q = -\lambda \, \text{grad} t = -\lambda \cdot \frac{\partial t}{\partial n} \qquad (2.1)$$

where $\partial t / \partial_n$ - partial derivative, n - normal to the isothermal surface. Here, the minus means that the vector of the heat flow directs towards the falling temperature. Coefficient of thermal conductivity λ numerically equals to the amount of heat that passes in unit time through unit area of the isothermal temperature surface at gradient equaled one.

Simultaneous solution of equations expressing the conservation of energy and Fourier law (2.1) leads to the heat transfer equation, which is valid for each phase of a thermodynamic system:

$$c \cdot \frac{dt}{dn} = \text{div}(\lambda \, \text{grad} t) + q_v \qquad (2.2)$$

where q_v - specific density of the internal heat sources, W/m³.

Equation (2.2) with the chosen nonlinear boundary conditions is usually attracted to the mathematical description of the problem of determining the temperature field in the system with moving boundaries phases.

For more than a hundred years, to solve heat conduction problems with moving boundaries there are used analytical methods, the essence of which is to obtain an explicit formula expressing the solution in terms of elementary or some special functions.

First solutions have been proposed by G. Lame and B. Clapeyron [8], J. von Neumann [9], and then outstanding thermophysics theorist I. Stephan [10]. He suggested determining the speed of the phase boundary movement by the heat removal of the latent heat of the phase change by equation

$$\lambda_s \cdot \frac{\partial ts}{\partial x} = \lambda_l \cdot \frac{\partial tl}{\partial x} - r \cdot \frac{dh}{d\tau} \qquad (2.3)$$

where the subscripts "s" and "l" refer respectively to the hardened body and liquid. Writing equation (2.3) shows that if during dτ, the solidification boundary will move to the interval dh, then the heat of phase change and the heat rejected from the liquid, have to be transferred through the hardened layer. It was assumed that the temperature fields in the liquid and solid phases are described by the heat conductivity equation with the appropriate thermal characteristics. The transition from one state to another is carried out on a smooth surface, on which, in turn, the solidification temperature is set. The distribution of heat occurs only along one spatial coordinate.

For spatial propagation of heat in the case of a stationary medium with constant physical properties without internal heat sources, Stefan's condition can be written as follows [11]:

$$r \cdot \frac{\partial F}{\partial \tau} + (\lambda_s \cdot \text{gradt}_s - \lambda_l \cdot \text{gradt}_l; \text{gradF})_{B\tau} = 0 \tag{2.4}$$

The difference between equations (2.3) and (2.4) is not only in the form of mathematical notation, but in the physical content. Writing equation (2.4) indicates that the propagation direction of heat flow may not coincide with the direction of advancement of the phase change surface.

Let n_t - vector that coplanar to the direction of heat flow; n_s -normal to the surface of the phase-change B_τ at some point. Then the scalar product in equation (2.4) can be written as

$$(\lambda_s \cdot \text{gradt}_s - \lambda_l \cdot \text{gradt}_l; \text{gradF}) = |\lambda_s \cdot \text{gradt}_s - \lambda_L \cdot \text{gradt}_l| \cdot |\text{gradF}| \cdot \cos(n_t, n_s) \tag{2.5}$$

The numerical value of $\cos(n_t, n_s)$ shows how many times the speed of advance of the phase-change surface decreases because of a mismatch of the propagation directions of heat flow and phase surface. For each point of the surface, value of $\cos(n_t, n_s)$ will be different.

Reduce (2.4) to dimensionless form. To do this, we rewrite it in the form of

$$r \cdot \frac{\partial F}{\partial \tau} + \left(\lambda_s \cdot \sum_{i=1}^{3} \frac{\partial t_s}{\partial x_i} - \lambda_l \cdot \sum_{i=1}^{3} \frac{\partial t_l}{\partial x_i} ; \sum_{i=1}^{3} \frac{\partial F}{\partial x_i} \right)_{B\tau} = 0 \tag{2.6}$$

and introduce scales of the length l^*, the coefficient of thermal conductivity λ^* and the temperature difference $\Delta t^* = t_1 - t_2$, where t_1 and t_2 − temperatures of characteristic points in the thermodynamic system. The corresponding dimensionless conversion factors [12, 13] have the form

$$\Delta \hat{t} = \Delta t/(\Delta t^*); \ \hat{\lambda} = \lambda/\lambda^*; \hat{x}_i = x_i/l^* \qquad (2.7)$$

where $\Delta t = t_1 - t_2$; "^" - a sign of the dimensionless quantity.

Substituting (2.7) in (2.6), we obtain

$$\frac{r}{c^*\Delta t^*\partial\frac{a^*\tau}{l^{*2}}} + (\hat{\lambda}_s \sum_{i=1}^{3}\frac{\partial\Delta\hat{t}_s}{\partial\hat{x}_i} - \hat{\lambda}_l \sum_{i=1}^{3}\frac{\partial\Delta\hat{t}_l}{\partial\hat{x}_i}; \sum_{i=1}^{3}\frac{\partial\hat{F}}{\partial\hat{x}})_{B\tau} = 0 \qquad (2.8)$$

$\hat{F} = F(\hat{x}, \hat{y}, \hat{z}) = 0$ - equation of the surface of the phase change in dimensionless coordinates B_τ; a^* – scale of the diffusion coefficient of heat (thermal diffusivity coefficient); c^* - scale of volumetric heat capacity.

In equation (2.8), three dimensionless conversion factors are involved (2.7), as well as complex

$$F_O = a^*\tau/l^{*2} \qquad (2.9)$$

called Fourier criterion and has the meaning of the dimensionless time, and the complex

$$\hat{r} = r/(c^*\Delta t^*) \qquad (2.10)$$

characterizing absorption of heat of phase change.

The complex $\hat{r}$ is a special case of thermal criterion of physico-chemical transformations

$$K = r'/(\Delta i) \qquad (2.11)$$

where, r'- heat of reaction; Δi- enthalpy pressure in the researched phase.

Criterion K is obtained by S.S. Kutateladze [13]. In this paper, a method of similarity theory has been applied to the processes of heat transfer during phase transformations. There were obtained criterial complexces depending on heat transfer during melting and solidification.

Among the fundamental research of heat transfer during melting and solidification, the work attributed by A.G. Tkachev [14] should be noted. He established a general view of criterial equation of solidif ication in the case of free convection of the liquid phase along the solid surface:

$$Nu = f\left(Gr, \text{Pr}, \text{Fo}_l, K_l, \frac{t_l - t_{cr}}{t_{cr} - t_0}, \frac{\lambda_l}{\lambda_s}, \frac{\rho_l}{\rho_s}, \frac{c_l}{c_s}\right);$$

$$Gr = gl^3\rho(t_l - t_{cr})/v_l; \quad \text{Pr} = v_l/a_l \qquad (2.12)$$

$$\text{Fo}_l = a_l\tau/l^2; K_l = r/[c_l(t_l - t_{cr})]$$

where g - acceleration due to gravity, 9.8 m/s^2, t_{cr} - cryoscopic temperature, °C.

Fourier criterion has some specific conditions for melting and solidification, when there is a continuous change in the size of solid body. Usually, in the research of nonstationary heat transfer theory, the conventional Fourier criterion depends mainly on the time. In problems associated with melting and solidification processes, a linear parameter in this criterion can vary considerably too.

To solve problems in which the heat transfer during solidification materials described various combinations of equations (2.2) , (2.3), (2.4), (2.12) and boundary conditions that take into account features of the occurrence of this process , many researchers use the variational method Bio [15], Goodman's integral method [16] , related to them , approximate methods , such as Vejnik method [6]. Fairly complete bibliography and a mathematical approach to the solution of the Stefan problem when using these methods are given by L.I. Rubinstein [11] and L.A. Kozdoba [17].

The main drawback of this group of methods for all their efficiency - the lack of criteria for choosing the approximating function of the temperature distribution, especially for bodies with axial and central symmetry, as well as for multi-dimensional bodies [18].

For problems of an applied nature, the method of linearizing substitutions is applied. It provides fairly simple formulas useful in engineering practice. Method reduces to the choice of substitution linking the desired temperature with a new function. So that the boundary value problem regarding a new function will have a linear boundary condition.

So, L.S. Leibenzon [19] proposed a method, the essence of which is to replace the true temperature distributions of each phase by their approximate counterparts satisfying the boundary conditions. These approximating functions are substituted into the Stefan condition and solve the given differential equation. As a result, the dependence of the freezing duration from the thickness of frozen layer is calculated. The method was successfully used in the analytical study of the processes of freezing and thawing of foods [20], and for the study of the kinetics of the freezing of the ice layer from liquid continuously flowing upon the unwetted surface [21, 22]. The main disadvantage of Leibenson's method is inability to take into account the specific heat capacity of the two phases [18].

Interesting and original solutions to the Stefan problem are also proposed in other publications [23, 24]. Paying tribute to the merits of each of these new proposals, it is impossible to choose any of them as a tool of practical calculation of a specific task with a large number of influencing parameters.

Because of the rapid development of computer technology, analytical methods in recent years have increasingly giving way to numerical methods for solving the Stefan problem. Nevertheless, analytical techniques are very powerful tool for making qualitative analysis of simple model problems.

A thorough review of the numerical schemes for solving the Stefan problem is presented in monographs A.A. Samarsky [25], L.I. Rubinstein [11], and N. Nikitenko [26]. Note through the numerical known methods, the most versatile is the grid method, otherwise known as the finite difference method, which allows solving both linear and nonlinear differential equations. This makes it possible to abandon the simplifying of mathematical model of solidification and melting. The solution can be obtained with any desired degree of accuracy [26]. The grid method has been extended also because when it is used there is no need to use or to specify analytical expressions for the equations of body boundaries, boundary conditions, transport coefficients, etc. [27]. Furthermore, a significant advantage of the method of nets is repeatability of identical mathematical operations when you use it. This creates very favorable conditions for the application of modern computing.

Difference methods based on the replacement of known functions by some other function, have great practical importance for the solution of the Stefan problem associated with melting and solidification of wet systems are [28]:

$$H(t) = \int_0^t [c(t) + r_v \delta(t - t_{cr})] dt, \tag{2.13}$$

where $H(t)$- the enthalpy per unit volume of material, J/м³, $\delta(t - t_{cr})$ - Dirac delta-function.

Conditions on the phase front (boundary) will be considered if the heat conductivity equation (2.1) is replaced by the equation

$$dH(t)/(d\tau) = \text{div}[\lambda(t) \cdot \text{grad } t] \tag{2.14}$$

Methods of analogies are very useful for the study of heat conductivity problems with phase-change phenomena. In 1922 academician Pavlovsky substantiated an electrodynamic analogy, laying the foundation for the modeling of physical fields in continuous media. In 1926, Soviet scientists Boris Gershgorin proposed method of the "electric grid" that greatly enhanced the ability of electrical analogy of fields. These ideas have been used for the analysis of heat conductivity problems based on the use of the analogy between electrical and thermal phenomena.

Limited applicability of analytical methods and the methods of analogy, difficulties in using computers and methods of computational mathematics, due to lack of qualified researchers, are the cause of numerous attempts to directly use of experimental results. Solutions resulting from correlation of experimental data, in the form of graphs, nomograms, and criteria equations allow us to judge the qualitative and, to a certain extent, the quantitative relationship of observed parameters of the investigated process. However, the experimental method can not explain the causes of the process movement exactly in the direction of what is observed in practice, and accurately substantiate the list of the selected parameters of the process. In addition, the use of obtained solutions is limited for their applicability.

Experience in solving problems associated with melting and solidification processes, showed that a preliminary analysis of a mathematical model by using the theory of similarity (the definition of set of physical criteria, each of which controls a specific behavior of physical phenomenon) and the subsequent application of numerical methods for their implementation on a computer, allow you to get a volume of information that can not be obtained yet by the analytical methods (both on the feasibility of the solution and the determining of the effect of all factors on the unknown function). However, the analytical methods in comparison with numerical methods, allow the obtaining more illustrative solutions, by which it is easy to analyze the impact of the selected factors on the result of decisions. Furthermore, in practice, it is usually considered a good result obtained with the error up to 10%, but the less accuracy is applicable too [29]. Therefore, research on solving the Stefan problem in relation to technological processes of solidification and melting is advantageously carried out on the basis of a synthesis of analytical and numerical methods.

However, the applicability of different calculation methods is defined by a number of features of the studied process of thin layer freezing (solidification) of materials on the cooled surfaces:

1) The number of the interacting bodies reaches, usually, three (cooled wall, the solidified substance, and unfrozen material);

2) The length of the zone at which the cooling, freezing and subcooling of product's layer are realized, is constrained by the finite dimensions of the drum freezers. It dictates the need to maintain constant initial and boundary conditions. In particular, the temperature distribution of the wall before putting a new doze of the product layer should be permanently constant during the time;

3) The process is accompanied by the latent heat of phase change that applies equally to substances with a certain phase change temperature, and the substances with a phase change in a certain temperature range;

4) For the implementation of a continuous process in time and space, it is necessary to provide a motion of interacting bodies relative feeding devices, in other words, the mathematical description of the process should take into account its dynamic nature;

5) In practice, the continuity of researched process is provided by applying a material layer or bulk material pieces of a given thickness on a cooled surface, i.e. there is carried a freezing of stereometric bodies of finite dimensions;

6) The technological process of freezing of liquids and pasty, minced-shaped products on the cooling surface is characterized by high heat flow. It occurs within short time intervals.

It should be noted that the analysis of thermal conditions of the material solidification on the moving cooled wall is not to look for a solution in all its particular complexity. In the present state of our knowledge such decision is impossible. It seems appropriate the providing of a general solution of the duration of the material's solidification, as a simple in form, convenient for practical use and with an error not exceeding permissible.

Material and cooling wall are selected for study as objects of the model development process. The material depositing on the cooled surface is under the refrigerating treatment, reaches the crystallization temperature, solidified, and finally, is subcooled to a predetermined final temperature. Interacting bodies (material and cooled wall) represent a complex system, separate areas of which are in different conditions: part of the product is subcooled, other part - cooled, third - solidified. Part of the wall in direct contact with the material, is heated relative to the other part that is not in contact with the material. Quantitative relationships, characterizing the mentioned condition, determine the duration of the phenomenon under study.

As the basic required variable, the duration of the solidification of the material to a predetermined temperature and the length of the solidification zone and exposure of the material layer to the desired final temperature (at a known constant velocity cooled wall) can be selected. Obviously, from the perspective of the designer in developing the theory of drum machines both approaches are entitled. However, the mechanisms of heat transfer in liquids and pasty, minced-shaped products are different.

At the solidification of liquid, the heat flux to the zone of the phase change is realized by convection. The heat transfer coefficient can be significant. For example, the heat transfer coefficient of water can reach 100-1,000 W/(m²·°K)), and it increases the duration of solidification.

When the paste solidification process is provided, the heat flux in the phase transition zone is not carried out by convection. It is realized by thermal conductivity through the layer of uncured material, i.e. ten times less intense. In addition, the pasta-shaped products, cooled in freezers, are very diverse on thermophysical properties.

These reasons explaine the need to develop a variety of physical-mathematical models (PMM) for two cases: the solidification of the liquid layer continuously impinging on a cooled surface, and hardening the layer of the paste-shaped product of a given thickness, deposited on a continuously moving cooled surface.

Difficulties of the development of PMM for the case of hardening paste and minced are compounded by the fact that most of the thermal characteristics of the processed materials are unstable and can vary in a certain range of values. Therefore, for engineering calculations is expedient to reduce the number of influencing parameters to a minimum. Such a formulation of the problem leads to the feasibility of the theory of similarity, as well as the theory of the active experiment planning with a view to the best of the experiment and processing the results. Due to the developed theory based on these principals, it can significantly improve the efficiency of research [30].

All of the above predetermined a goal of this book: to offer engineering design methods for calculating of construction and mode characteristics of drum machines on the basis of studies of heat transfer during freezing of paste, minced-shaped products and liquids.

2.2. Heat transfer of thin layer freezing of pastor and minced-shaped products

To construct the process PMM of a thin freezing (solidification) of pastor and minced-shaped material on a moving cooled wall, there is needed a combination of factors and patterns that determine the actual temperature field and choose the most important from them. This should take into account the existing experience of formation of physical concepts and mathematical description of the process under study.

To date, in the literature, several approximate analytical and numerical methods for solving two- and three-dimensional problems related to the solidification process are published. A brief overview of these methods can be found in [31, 32].

The particular note is the works [33, 37]. In addition, R. Siegel [35] performed by the method of conformal mappings, the analysis of two-dimensional solidification of the ingot, cooled and drained continuously vertically downwardly from a mold having the parallel walls of the finite length. With this method, the author developed Laplace's equation, which describes the transfer of energy in the region occupied by the solidified material. The advantage of this method is that it can be applied to bodies of arbitrary shape. However, the spread of the method to the problem associated with the solution of elliptic equations (for the case of the transfer by heat conduction and convection) may be associated with considerable difficulties [36].

Assessing the impact of the axial component of the heat flux on the metal solidification in continuous casting of the flat ingot is discussed in [37]. Assuming that on the cool surface is realized the boundary condition of type II, and the temperature of the liquid phase is equal to the crystallization temperature t_{cr}, the author by the analysis of the self-similar solutions, obtains the condition for the validity of the approximate solution of the problem without considering the axial component of the flow:

$$r \cdot Pe/(2q_{out}) \geq 1,$$

where $Pe = c_{mcr} v^* x_0 / \lambda_{mcr}$; λ_{mcr}, c_{mcr} - thermal conductivity and specific heat of the metal at the crystallization temperature t_{cr}, respectively; v^* - speed of movement; x_0 - half the thickness of the ingot; q_{out} - heat flux density on the outer surface of the solidified ingot; r – latent heat of crystallization of the metal.

This condition gives an indication of the structure-form process flow of a continuous ingot. In addition, the resulting solution can be used as a test to verify the numerical algorithms for the calculation of continuous ingot solidification in the form of flat shape with physical characteristics that depended from the metal temperature.

A significant drawback of the approach is that the body undergoes a phase change, must be initially at the crystallization temperature, whereas in practice it is difficult to provide conditions which exclude initial overheating (or supercooling), especially for the body of substantial size.

Significantly more similar in formulation of the task to the problem of interest to us is the work of Z.M. Komladze [38]. The author examines the process of two-dimensional freezing of the viscous wet material on the cooling surface in their relative motion. Assuming that the heat flow in the direction of motion is negligible, the one dimensional problem is solved. The resulting solution expresses the relationship of factors determining the thickness of the frozen layer h:

$$h = \{2 \cdot \lambda_s \cdot t_{inm} \cdot y \cdot [1 - p_{dr} \cdot y/(2 \cdot \Omega)]/[v^* \cdot (r \cdot W \cdot \omega \cdot \rho_1 + \lambda_l \cdot t_{inw}/a_l)]\}^{-1/2}$$

where v^* - velocity of the frozen product: p_{dr} - coefficient characterizing the structural features of the freezing drum; Ω - the width of the supplying camera; t_{inm} and t_{inw} - initial temperature of the material and the wall, respectively.

As pointed out by the author himself, the obtained analytical dependences can be recommended for the calculation of the freezing process of the wet viscous material on a moving cooled surface at relatively low values of speed (not more than 0.005 m / s).

In [39] there is an attempt to investigate the possibility of optimization of some indicators of ingot quality based on its calculation of the temperature field (in the two-dimensional approximation) in a cylindrical coordinate system. To solve the heat equation, which describes the transfer of heat in the body of the ingot, the author used the method proposed by C.L. Kamenomostskaja [28]. Conditions on the "border ingot - cooling surface" is defined in three ways: to define the experimental data; theoretically assumed to be selected in the search for the optimal choices: be recovered from the known profile of the solidification front. It should be noted some awkwardness and subjective evaluations of the proposed system of "quality of the ingot." Furthermore, it is obvious that the solution of the problem by setting the temperature on the cooling surface, does not consider the influence of thermal properties and dimensions of the cooling body.

Analysis of the above-mentioned works allows claiming that the previously proposed formulas for the calculation of the duration of the solidification of the material layer of a given thickness are obtained by the introduction of numerous simplifying assumptions, with the adoption of different boundary conditions and methods that taking into account the effect of thermophysical properties of the interacting bodies.

In our application of, in the study of complex physical processes, which are characterized by a large number of influencing factors/varaibles, and, in particular, the process of solidification of the material layer on a moving cooling surface, it is not enough substantiated the admissibility of certain simplifications; there is no assessment of the relative influence of factors on the final outcome of studies (productivity, quality, material, time, freezing, etc.).

The suggested analysis of earlier designs of the continuous action devices shows that not addressed the issue of the development of such DF that could be used to freeze a number of products with different physical properties. Therefore it is necessary to evaluate the degree of the relative influence of thermophysical properties of frozen materials on the character of the process of interest.

The results obtained by solving PMM of the freezing of the liquid (water) can not be used to calculate the solidification duration and supercooling of layer of pastors, minced-shaped products of certain predetermined thickness to the specific final temperature. This is primarily due to the fact that at the thin layer solidofication of frozen foods, the heat transfer occurs mainly by conduction, whereas for freezing liquids the main mechanism of heat transfer is convection. There are several other differences.

Consider the main features of the process of thin-layer freezing of pastor, minced-shaped products and reasonable assumptions needed to construct the computational model.

1. In a rigorous theoretical approach to the study of the formation of a frozen layer of the product, there is required an analysis of all the complex phenomena occurring in the layer: thermodynamic, hydrodynamic, diffusion, thermal. In this paper we study the process described in the model, which takes into account only the heat transfer by conduction in the interacting bodies (materials, wall). It is advisable to carry out the process under study so that on a cooled wall the thin layer of the product will be deposited. It contacts with the ambient air at the opposite side. Thus heat leakage will be realized into the phase change zone by conductivity through a layer of small thickness.

2. The real three-dimensional temperature field, however, in some cases, can be legally considered as a two-dimensional or even one-dimensional approximation. In the study of thin-layer freezing of the product on a cooled wall, the heat flux through the lateral surfaces of the layer of the product and the wall (q_{ss}) can be neglected, as in the real conditions, the densities of heat flow through the butts are two to three orders of magnitude lower than in the radial direction (q_y) and coinciding with direction of the wall and product movement (q_x), i.e., $q_y \gg q_{ss}$, $q_x \gg q_{ss}$.

Solution of the problem by assuming propagation of heat in one direction carries considerable inaccuracy, since in reality the cooling treatment of materials (including nutritional) never occurs on the endless cylindrical or flat surfaces. Rather, in practice, the temperature field is always differs significantly from the one-dimensional [40]. When studying the process of freezing a layer of material in the continuous DF, the heat transfer must be taken into account. The heat is supplied by the applied material that is put along the cooled wall. In other words, it needs to consider the axial component of the heat flux. This phenomenon, of course, affects the nature of the temperature distribution in the wall, which is "heating" in comparison with the wall, on the surface of which the material is missing.

On the other hand, the limited extent of the zone in which the cooling, freezing and subcooling of the material layer are realized, dictates (due to the finite dimensions of the drum machine) the need to maintain constant initial and boundary conditions (in particular, the temperature field in the wall prior to the application of the material layer must be permanently over time). By the ratio between the length of the "thermal inertia" zone of a wall and the characteristic geometric size of the device (the circumference of the drum) it can indirectly judge the heat flux propagating along the direction of movement of the wall with the product, i.e. on the admissibility of the application of the one-dimensional or two-dimensional description of the researched process.

Calculations show that for the possible modes of operation of DF designed for freezing and subcooling of minced-shaped products of sea fishing, the length of the "thermal inertia" zone of a wall may reach (30-50)% of the circumference of the drum-evaporator. Thus, for a detailed study of the process of freezing the product layer on a cooled wall of the drum, it is expedient to use a two-dimensional approximation for describing the process.

3. To date, there are no reliable data on the thermal characteristics of pastors and minced-shaped products of sea fishing. For other types of foods, it should be noted insufficient accuracy and especially the limited thermal measurements of fundamental character at temperatures considered in this work from -40°C to 40°C.

This is due to the complexity of the structure of specified materials (solids of varied structure and liquid solutions of different concentrations, in which gas inclusions may be located). Their thermal characteristics depend not only on the temperature, density and moisture content, but also on the chemical composition and structure of the product and the experimental conditions [41, 4].

It has been established [29] that for technical computing is permissible to use thermal characteristics of foods that have been found with an error of (5-8)%. However, in practice the less accuracy is usually available.

Given the above and focusing on computational and experimental data in [4], we establish the possible range of change of thermal characteristics. And specific conditions in each of the interacting bodies (initial material, the hardened layer, and wall) we reflect, by the introduction for each of them, the constant characteristics (in particular, the thermal conductivity and volumetric heat capacity).

Since foods are complex heterogeneous systems [42], the release of the latent heat of phase change occurs in a certain temperature range 2Δ, comprising cryoscopic temperature. From a physical point of view, this is equivalent to assuming that there is no the sharp phase boundary. This "blurring" of the boundaries of the phase change is also advisable from a mathematical point of view, because from the formulation of the problem, there is formally excluded the Stefan condition at the interface. So the problem reduces to the solution of a single heat transfer equation with variable coefficients for the interacting bodies [43].

4. conditions in the interacting bodies, according to the accepted hypothesis are the determining For most technical problems, it may neglect curvature of the product layer and the evaporator wall, considering a wall as a planar due to low thickness of both two layers compared with the radius of the evaporator.

5. Studying any process is to develop an engineering method of calculating machines that implement this process. For this purpose, it seems the steady process is eligible in the continuous apparatus. It meets the practical modes of the device. Therefore we use the quasi-stationary formulation of the problem.

6. Thermal factor characterizing the process of solidification of minced or paste on the cooled wall. The boundary conditions of the problem are defined by the laws of heat transfer between: the product layer and the environment, between the product layer and the wall, the wall and the cooling medium, and the thermal state of the wall and a product prior to the application of the latter on the cooling surface and after its removal from the wall. For the system "the material - a wall" that could be studied as a single body, we suppose that the thermal resistance of the contact of "material – wall" is equal to zero. In other words, on the surface of contact between the two bodies, a condition of equality of temperature (the continuity of the temperature field) and the heat flux density (energy conservation) will be made.

For further construction of the model it is advisable to carry out using the scheme shown in Fig. 2.3. On the plate of thickness R, the product is fed at a predetermined temperature t_{in}. It is formed into a layer of thickness h. The lower surface of the wall is cooled by the refrigerant/coolant. The product with the wall is moved with the same velocity V relatively to the feed point of the product. The frozen product layer that is supercooled to a final temperature t_f along the length l is removed.

Temperature fields of interacting bodies in the model satisfy the equation of heat conduction with heat capacity c_{eff} (t), taking into account the release of latent heat.

It is assumed that the coordinate system with respect to which the product and the wall are moved in the direction of the axis of abscissas, is rigidly connected to the feeder - a device for applying the product layer of a predetermined thickness on the surface of the drum-evaporator.

The boundary conditions of III-d kind are given on the refrigerant side (or chilled water) and the environment side. By other words, a dependence of a density (W/m²) of the heat flux transmitted to outer surface of the product and inner surface of a wall is given. The dependence is caused by the temperature difference between the body surface and the medium (air or refrigerant/coolant). This heat flux is supplied by the thermal conductivity.

The boundary conditions of III-d kind are given on the refrigerant side (or chilled water) and the environment side. By other words, a dependence of a density (W/m2) of the heat flux transmitted to outer surface of the product and inner surface of a wall is given. The dependence is caused by the temperature difference between the body surface

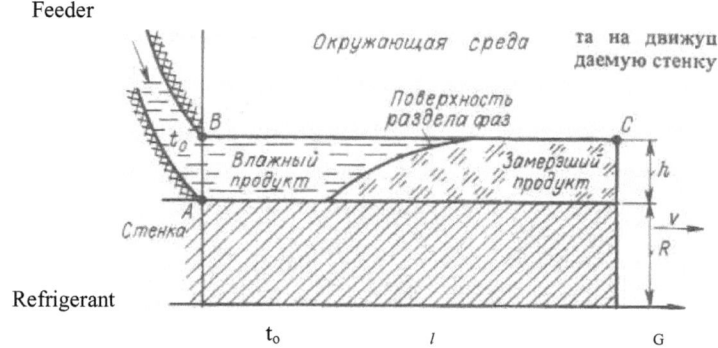

Fig.2.3 The scheme of applying a layer of paste-shaped product
on a moving cooling wall

and the medium (air or refrigerant/coolant). This heat flux is supplied by the thermal conductivity.

In the section x = *l* the homogeneous boundary condition of type II (the distribution of the heat flux on the surface of the body as a function of the coordinates) is given, i.e., condition of the so-called "thermal insulation". Last assumption is acceptable because *l* is not constant, but has the finite value depending on the temperature and technological modes of the drum unit.

In the cross-section of x = 0 along the wall thickness (up to the point of supplying the product on the cooled surface), a linear temperature distribution is defined. It determines by the temperatures of the refrigerant and the environment. The temperature along a thickness of the product layer is assumed constant and equal to the initial temperature of the product.

In this setting, solution of the problem is due to the need to consider the set of the realizable in practice values of design and operational parameters of the researched process. This, in turn, causes the execution of a large amount of calculations. So the use of a computer is a necessary condition to obtain a quantitative decision-making.

Once the physical formulation of the problem is defined, there is conducted its study by computer experiment [44].

Computer experiment consists of several stages. The first step is to compile equations of the problem expressing in quantitative form a general idea of the physical mechanism of the process. They are based on the analysis of the process as a particular application of the fundamental principles of physics. In most cases they are the form of differential (integral, integro-differential) equations.

Since the studied process is quite complicated and it can not be investigated on the basis of only one physical law, there is a need to consider various aspects of the model and to attract different physical laws. Therefore, usually, the overall process is determined by the system of equations.

In addition to the basic equations there are written boundary conditions: a set of constant parameters characterizing the geometric and physical properties of the system that are essential for the process as well as conditions for uniqueness.

After mathematical model is made, it is necessary to determine the correctness of its formulation (existence of a solution, its uniqueness, does it continuously depend on the boundary conditions). However, in practice for many applications it is impossible to rigorously prove theorems of existence and uniqueness. So there are used some "illegal" mathematical techniques that do not have a precise mathematical justification [45].

In the second stage of computational experiment, the selection of the computation algorithm is realized. In a broad sense, the algorithm refers to the exact prescription that specifies the computational process, starting from an arbitrary initial datum and aimed for obtaining results which completely defined by this initial data [46]. In a narrow sense, computational algorithms are the sequence of arithmetic and logical operations, by which the mathematical problem solution is solving [44].

Computational algorithm focused on use of the modern computers must meet the following requirements: 1) to give a solution of the original problem with a given accuracy after a finite number of actions; 2) implement the decisions of the problem by spending the least possible computer time; 3) to ensure the absence of an emergency stop (abend) of computers during the calculations; 4) be sustainable (in the process of calculation the rounding errors should not be accumulated). For more detailed information about this phase of computational experiment, it can be recommended [47].

In the third stage, the programming of computational algorithm for a computer is organized. This issue is devoted to a huge amount of work. Given the specificity of this study, the greatest interest is the work [48].

The fourth stage involves performing calculations on a computer, and the fifth - with the analysis of the numerical results and the subsequent refinement of the mathematical model.

From the standpoint of saving computer time and the practical value of the information received, the organization and planning of the last two stages of the computational experiment are important. So, just before the start of computational experiment, the question of the scope and methods of processing (convolution) output data should be carefully considered. Obviously, in the study of any process, experimenter faces to accommodate a large number of variables (factors[1]), and accordingly, the solving of the multiextremal problem.

In view of the above, you can formulate the main problems in the mathematical description of the process of thin-layer freezing of pastor and minced-shaped products:

- to develop a refined mathematical model of the process under investigation in the two-dimensional approximation, allowing the use of the corresponding calculation method in the design of unified series of DF for freezing of products with

---

[1] Factor is a variable that takes a certain numerical values and corresponding to one of the possible ways to influence on the object of study.

different thermal characteristics in a wide range of changes in external conditions;

- to determine the relative influence of operating and design characteristics of the device at the character of a continuous process of freezing a thin product by computer simulation;

- to verify analytical dependences, including the essential parameters of the test process and allowing you to develop rational design of DF.

CHAPTER 3: THEORY AND CALCULATION OF FREEZING AND SUPERCOOLING OF THIN LAYER OF PASTED-MINCED PRODUCTS IN DRUM FREEZER

3.1. Mathematical model of thin layer freezing of pastor, minced-shaped products

In the study of complex problems such as the freezing of the process of thin pasty, minced-shaped products on a moving cooled wall, very important is the choice of a mathematical model of the object being studied. It should be sufficiently accurate and adequate reality, however, is not overly complex, suitable for finding the optimal variant the computer. These requirements are to some extent contradictory, so at the design of a mathematical model that is suitable for the present study, in some cases, have to resort to compromise solutions.

We carry the development and solution of the mathematical model according to the plan of computational experiment (see section 2.3), using the provisions of the process physical model set out in the same place.

For temperature fields of interacting bodies (product-wall), in the general case, we have the energy equation

$$dH(t)/d\tau = div[\lambda(t) \cdot gradt], \qquad (3.1)$$

where

$$H(t) = \int_0^l [c(t) + r_v W \omega \delta(t - t_{cr})]dt \qquad (3.2)$$

Equation (3.2) differs from the corresponding equation (2.13) by entering, into the second term, the integrand factor $W \cdot \omega$ that takes into account the fact that, firstly, the product consists of solids and liquids, and, secondly, not all of the liquid crystallizes.

Equation (3.1) includes both the heat transfer equation (2.2) that is valid for each interacting body, and equation (2.4) at the phase boundary [43, 49].

Equation (3.1) can be simplified by substituting the expression for H(t) (3.2) and given the fact that the promotion of the wall with the product is carried out at a constant velocity $v=dx/d\tau$ along the x-axis of the coordinate system rigidly fixed with respect to the feed point of the product (Fig. 2.4). Then

$$v[c_v(t) + r_v W \omega \delta(t - t_{cr}, \Delta)]\frac{dt}{dx} = \frac{\partial}{\partial x}\left(\lambda(t)\frac{\partial t}{\partial x}\right) + \frac{\partial}{\partial y}\left(\lambda(t)\frac{\partial t}{\partial y}\right) \qquad (3.3)$$

where $\delta(t - t_{cr}, \Delta)$ - delta-function, allowing to enter "smoothed" or effective heat capacity:

$$c_{eff}(t) = c_v(t) + r_V \cdot W \cdot \omega \cdot \delta(t - t_{cr}, \Delta) \qquad (3.4)$$

Here Δ - the semi-temperature interval on which $\delta(t - t_{cr}, \Delta)$ is non-zero.

For the smoothed heat capacity one can use a dependence that is analogous to that proposed by A.A. Samarsky [49]:

$$c_{eff}(t) = \begin{cases} r_v \cdot W \cdot \omega/(2\Delta) + \end{cases} \begin{matrix} c_s, t < t_{cr} - \Delta, R < y \le R + h; \\ (c_s + c_l)/2; t_{cr} - \Delta < t < t_{cr} + \Delta; R < y \le R + h; \\ c_l, t > t_{cr} + \Delta, R < y \le R + h; \\ c_p, 0 \le y \le R \end{matrix} \qquad (3.5)$$

Thus, equation (3.3) includes both the heat equation (3.1) that is valid for each interacting body, and the condition (2.4) at the boundary of phase changing [50, 49].

Conditions at the boundaries of the interacting bodies (wet product, frozen product, the wall) can be written as follows:
on the segment AB (see Fig. 2.4) ($x = 0, R < y \le R + h$) the product temperature is constant and equal along its thickness

$$t(0, y) = t_{in} \qquad (3.6)$$

on the segment OA ($x = 0, \ 0 \le y \le R$) the temperature distribution across the wall thickness before the supplying of the product layer is linear

$$t(0, y) = (t_{env}/(1/\alpha_0 + 1/\alpha_{env} + R/\lambda_p)) \cdot (y/\lambda_p + 1/\alpha_0) \qquad (3.7)$$

where o, env, p - the symbols relating to the refrigerant, environment and wall, respectively;
on the segment OG ($0 < x < l, y = 0$), the convective heat transfer between the inner wall surface and refrigerant is characterized by conditions of type III

$$\alpha_0 \cdot (t_{inp} - t_0) = (\partial t/ \partial y)_{y=0} \cdot \lambda_p \qquad (3.8)$$

where t_{inp} - temperature on the wall inner surface;
on the segment BC ($x = l, 0 \le y \le R + h$) the convective heat transfer between the outer surface of product layer and the environment, is characterized by the condition of type III

$$\alpha_{env} \cdot (t_{env} - t_{out}) = (\partial t/ \partial y)_{y=R+h} \cdot \lambda_{out} \qquad (3.9)$$

where t_{out} - temperature on the outer surface of the wall; λ_{out} - thermal conductivity of the product on the outer surface of the layer;

on the segment CG $(x = l, 0 \leq y \leq R + h)$, due to the smallness of the heat flux along the x-axis at large distances from the origin, the following condition is realized

$$\partial t / \partial y = 0 \qquad (3.10)$$

The process of freezing and subcooling of the product layer on a moving cooled wall in two-dimensional space on a closed rectangular region is described by the equations (3.3) - (3.10).

It is obvious that all the physical dimensional quantities/variables appearing in the system of equations (3.3) - (3.10) cannot make an infinite number of different values in the real world. These values lay in certain intervals, the boundaries of which were selected based on experience operating of freezers, analysis of published scientific, technical, regulatory and technological literature of thermo-physical characteristics of the raw material and the wall material, design, and operational characteristics of the freezers.

Thus, according to data of G.B. Chizhov [29], the temperature of crystallization of most foods is in the range of $(-2 - 0)^0C$. This interval was used in the calculations.

Calculations made by V.B. Rzhevskaya [21] showed that the film coefficient in drum ice generators producing flake ice, can reach values of $(1.1-29.0) \cdot 10^3$ W/$(m^2 \cdot K)$, which are taken as the starting point in the mathematical model.

The thickness of the walls of vessels under pressure, according to the current rules is selected on the basis that the ambient temperature is 50°C. Then the thickness of the walls of the drum-evaporator made of corrosion-resistant steel SS316L can reach 0,012 m, and in the case of alloys aluminum brand AMg5 -0.016 m, instead 0,008-0,012 m corresponding to the drum flake ice generators used at the moment. This explains the choice of the permissible range of variation of the wall thickness from 0.008 to 0.016 m.

The reasons for the choice of admissible intervals for the rest of the physical characteristics included in the original system of equations can be similarly explained.

The numerical values of dimensional physical variables characterizing the process under study and introduced in the equations (3.3) - (3.10) are given in Table. 3.1.

In the proposed mathematical model there were introduced 19 physical quantities/variables. It should be noted that to find hidden relationships between variables in this case is very difficult, so it is very valuable to use the methods of the theory of similarity, which is in accordance with modern ideas can be called a theory of generalized variables [51]. Application of this theory is advisable for several reasons.

In real conditions, the influence of individual factors (many links that exist between the material object and the environment), represented by different values, is manifested not separately, but together. Therefore it is necessary to consider not individual values, but their set or complexes having a definite physical meaning. Methods of the theory of similarity allow, based on the analysis of differential equations and boundary conditions, to find these complexes. Furthermore, the transition from the normal physical quantities to the complex type variables, which are composed of the same quantities, allows reducing the number of variables.

Thirdly, a predetermined value of set of variables can be obtained as the result of many different combinations included in the complex. This means that, when considering the problem in the new variables, there is studied not an isolated case, but the countless of different occasions, united by some common properties.

The rules of transition from differential equations to the expressions in the final form are described in detail in [51]. In order to formulate the equations (3.3) - (3.10) to a dimensionless form, we choose as the scale for: the temperature difference t^*_{cr} -t^*_o=29°C, the length scale - the thickness of the product layer $h^* = 3{,}0 \cdot 10^{-3}$ m , the scale of the specific heat - the heat capacity of the original product $c_l = 3{,}0 \cdot 10^6$ J/(m³·K), the scale of thermal conductivity - thermal conductivity of the intial material $\lambda_l = 0.5$ W/(m·K).

We introduce the dimensionless temperature

$$\theta = (t - t^*_0)/(t^*_{cr} - t^*_0) \tag{3.11}$$

Next, we introduce the dimensionless variables

$$\hat{x} = x/h^*, \hat{y} = y/h^*, \hat{R} = R/h^*, \hat{\lambda} = \lambda/\lambda_l^*$$
$$\hat{c} = c/c_l, \hat{\Delta} = \Delta/(t_{cr}{}^* - t_0{}^*) \tag{3.12}$$

It is necessary to take into account that

$$\frac{\partial t}{\partial x} = \frac{t_{cr}{}^* - t_0{}^*}{h^*}\frac{\partial\theta}{\partial\hat{x}}, \frac{\partial t}{\partial y} = \frac{t_{cr}{}^* - t_0{}^*}{h^*}\frac{\partial\theta}{\partial\hat{y}}$$

38

After making a number of simple transformations, we obtain a system of dimensionless equations:

$$\hat{c}_{eff}\frac{\partial\theta}{\partial\hat{x}} = \frac{1}{Z_v}\left(\frac{\partial}{\partial\hat{x}}\hat{\lambda}\frac{\partial\theta}{\partial\hat{x}} + \frac{\partial}{\partial\hat{y}}\hat{\lambda}\frac{\partial\theta}{\partial\hat{y}}\right), 0 < \hat{x} < \xi, 0 < \hat{y} < \hat{R} + 1 \qquad (3.13)$$

$$\hat{c}_{eff} = \begin{cases} \hat{c}_{pl}, 0 < 1 - \hat{\Delta}, \hat{R} < \hat{y} < \hat{R} + 1 \\ \hat{r}_v/2\hat{\Delta} + (\hat{c}_{pl} + 1)/2, 1 - \hat{\Delta} \le 0 \le 1 + \hat{\Delta}, \hat{R} < \hat{y} < \hat{R} + 1 \\ 1, \theta > 1 + \hat{\Delta}, \hat{R} < \hat{y} < \hat{R} + 1 \\ \hat{c}_{pl}, 0 \le \hat{y} \le \hat{R} \end{cases} \qquad (3.14)$$

The boundary conditions are:
on the segment AB (Fig. 2.4)

$$\theta = \theta_{in}, \hat{x} = 0, \hat{R} < y < \hat{R} + 1 \qquad (3.15)$$

on the segment OA

$$\theta = \theta_{env}/[(1/e_1 + 1/e_2 + \hat{R}/\hat{\lambda}_{pl})](\hat{y}/\hat{\lambda}_{pl} + 1/e_1), \hat{x} = 0, 0 < \hat{y} < \hat{R} \qquad (3.16)$$

on the segment OG

$$(e_1/\hat{\lambda}_{pl})\theta_{out} = (\partial\theta/\partial\hat{y})_{\hat{y}=0}, 0 < \hat{x} < \xi, \hat{y} = 0 \qquad (3.17)$$

on the segment BC

$$(e_2/\hat{\lambda}_{out})(\theta_{env} - \theta_{out}) = (\partial\theta/\partial\hat{y})_{\hat{y}=R+1}, 0 < \hat{x} < \xi, \hat{y} = \hat{R} + 1 \qquad (3.18)$$

on the segment CG ξ

$$\partial\theta/\partial\hat{x} = 0, \hat{x} = \xi, 0 \le \hat{y} \le \hat{R} + 1 \qquad (3.19)$$

Table 3.1. The numerical values of physical variables characterizing the freezing process of thin layer

Physical variable	Dimension	Unit	Numerical values		
			Low level	Main level	High level
			-	*	+
Linear velocity	m/s	V	$1.85 \cdot 10^2$	-	$9.26 \cdot 10^2$
Thickness of evaporator wall	m	R	$8.0 \cdot 10^{-3}$	-	$16.0 \cdot 10^{-3}$
Thickness of product layer	m	h	$1.0 \cdot 10^{-3}$	$3.0 \cdot 10^{-3}$	$5.0 \cdot 10^{-3}$
Thermal conductivity of wall	W/(м·K)	λ_p	14.7	-	119.0
Heat capacity coefficient of evaporator wall	J/(м³·K)	c_p	$25.2 \cdot 10^6$	-	$72.0 \cdot 10^6$
Thermal conductivity of initial product	W/(м·K)	λ_l	0.30	0.50	0.70
Heat capacity coefficient of initial product	J/(м³·K)	c_l	$3 \cdot 10^6$	$3.6 \cdot 10^6$	$4.2 \cdot 10^6$
Thermal conductivity of frozen product	W/(м·K)	λ_{fr}	1.2	-	1.7
Heat capacity coefficient of frozen product	J/(м³·K)	c_{fr}	$1.5 \cdot 10^6$	-	$1.7 \cdot 10^6$
Initial temperature of product	°C	t_{in}	5.0	-	40.0
Final temperature of product	°C	t_f	-24.0	-	-3.0
Crystallization temperature of product	°C	t_{cr}	-2.0	-1.0	0.0
Evaporating temperature of refrigerant	°C	t_0	-40.0	-30.0	-20.0
Film coefficient of refrigerant	W/ (м²·K)	α_0	$1.1 \cdot 10^6$	-	$29.0 \cdot 10^6$
Temperature of environment	°C	t_{out}	5.0	-	20.0
Film coefficient from environment to product	W/(м²·K)	α_{out}	5.0	-	15.0
Latent heat of water crystallization	J/м³	r_V		$334.0 \cdot 10^6$	
Moisture content	%	W	60.0	-	85.0
	%	W	60.0	-	90.0

The main unknown variable is the dimensionless length $\xi = l/h^*$ - extent of the zone of freezing and sub-cooling of the material, placed on the cooled wall, to a predetermined temperature at which there is the condition

$$\theta_{out} - \theta_f < 0 \qquad (3.20)$$

Thus, the problem (3.3) - (3.10) reduces to the solution of the elliptic equation (3.13) with the boundary conditions (3.15) - (3.19) in a rectangular domain.

Table 3.2. Similarity criteria, dimensionless transformation factors and their varying intervals

Designation	Formula	Variation interval		
		Low level	Main level	High level
		-	*	+
Z_v	$v \cdot h^* \cdot c_i^* / \lambda_i^*$	400.00	1,200.00	2,000.00
θ_f	$(t_f - t_0^*)/(t_{cr} - t_0^*)$	0.21	0.57	0.93
θ_{in}	$(t_{in} - t_0^*)/(t_{cr} - t_0^*)$	1.21	1.82	2.42
$\hat{R}$	R/h^*	2.67	4.00	5.33
e_1	$\alpha_0 \cdot h^* \cdot \lambda_0^*$	6.60	90.30	174.00
$\hat{r}_v$	$r_v \cdot W \cdot \omega/[c_i^* \cdot (t_{cr} - t_0^*)]$	1.15	1.80	2.45
θ_{out}	$(t_{out} - t_0^*)/(t_{cr} - t_0^*)$	1.21	1.53	1.85
e_2	$\alpha_{out} \cdot h^* / \lambda_i^*$	0.03	0.06	0.09
$\hat{c}_{sl}$	c_s/c_i^*	0.42	0.45	0.47
$\hat{c}_{pl}$	c_p/c_i^*	0.70	10.35	20.00
$\hat{\lambda}_{sl}$	λ_s/λ_i^*	2.40	2.90	3.40
$\hat{\lambda}_{pl}$	λ_p/λ_i^*	29.40	133.70	238.00

Similarity criteria and dimensionless factors resulted due to the conversion of the original system of equations (3.3) - (3.10) in dimensionless form (3.13) - (3.19) are presented in Table. 4.2.

The intervals of varying of influencing parameters (Table. 3.2) are calculated by using the data in Table. 3.1.

For the entire course of further discussion and evaluation of the final result should be clear to the extent possible, the physical meaning of dimensionless complexes (Table. 3.2).

The meaning of the most of dimensionless complexes is obvious: θ_f and θ_{in} - final and initial temperature of the material deposited on the cooled wall; $\hat{c}_{pl}$ and $\hat{\lambda}_{pl}$ volumetric heat capacity and thermal conductivity coefficient of the wall; $\hat{c}_{sl}$ and $\hat{\lambda}_{sl}$ volumetric heat capacity and thermal conductivity coefficient of the solidified material; θ_{env} - the ambient temperature; $\hat{R}$ - wall thickness. Disclosure of the physical nature of the dimensionless complexes Z_v, $\hat{r}_v$, e_1, e_2 is difficult for several reasons.

At the moment, the similarity theory does not answer the question about the number of possible combinations of dimensional characteristics included in the description of the dimensionless physical process, and the form of these combinations. In addition, it is not clear what criteria, for many interacting bodies (in composite bodies), are suitable for the description of the physical process and how much they are required for a given error in the determination of the chosen main variable [52, 53].

By way of writing (the name and location of the order in the formula of dimensional variables) these criteria resemble respectively criteria Pe, K, Nu. However, the introduced here criteria are different from the well-known criteria due to the physical content of their constituent physical dimensional variables.

Thus, Z_v - the criterion that takes into account the movement of the wall with material in the coordinate system rigidly secured to place at which the product is applied on a cooled surface, is different from Pe, reflecting the balance between the convective heat transfer and thermal conductivity [54]. It is explained, first of all, by the fact that in our study there is no real, physical convection of cooled material relative to the wall.

$\hat{r}_v$ criterion, taking into account the release of the latent heat of phase changing, chara cterizes the ratio between the amount of heat withdrawn during the actual process of the phase transition, and the amount of heat that would be required to withdraw the product from the cryoscopy cooling temperature t_{cr} to the coolant temperature t_0. It does not allow considering $\hat{r}_v$ as a special case of the criterion of physical and chemical transformations K [13].

Criterion e_1, taking into account the film coefficient from the coolant, characterizes the relationship between the resistance of the applied layer of the product and the resistance in the contact zone "refrigerant - the inner surface of the wall"; criterion e_2, taking into account the film coefficient from the environment (for example, air), characterizes the relationship between the thermal resistances of the liquid layer and the contact between the environment and the outer surface of the material layer. It is clear that e_1 and e_2 are not equivalent to the Nusselt number for the appropriate boundary conditions.

The system of differential equations (3.13) - (3.20) allows, under certain parameter values of Z_v, $\hat{r}_v$, e_1, e_2, θ_f, θ_{in}, θ_{env}, $\hat{c}_{pl}$ and $\hat{\lambda}_{pl}$, $\hat{c}_{sl}$ and $\hat{\lambda}_{sl}$, $\hat{R}$, to calculate the temperature distribution in the interacting bodies and determine the extent of the zone ξ of cooling and freezing of the material on a cooled wall to the desired final temperature. In the second stage of computational experiment, the computational algorithm is selected.

As shown by the dimensional analysis, the temperature gradient in the y-direction is much larger than in the x-direction. In addition, the "smearing" of δ-function is carried out on a sufficiently small temperature range $2\hat{\Delta}$. Moreover, according to the recommendations of [49], two-three grid nodes should be accounted in this interval. Both factors lead to the fact that, firstly, the number of nodes of the finite-difference grid coordinate y must be high (at least 50 points are accounted on the product layer) and, the second, the step of a finite difference grid for the coordinate x may is taken to be substantially greater than the pitch in the y coordinate.

On the choice of the interval of "smearing" 2Δ, there is necessary to make the following observations.

In accordance with the recommendations of [49], the choice of this interval has been automated and constantly covered two countable points. Therefore its minimum $(2\hat{\Delta})_{-}$ and maximum $(2\hat{\Delta})_{+}$ values may be calculated by the following formulas:

$$(2\hat{\Delta}) \approx 2(\theta_{in}{}^{-} - \theta_f{}^{+})/50; \quad (2\hat{\Delta}) \approx (\theta_{in}{}^{-} - \theta_f{}^{+})/50$$

Using the data from Table.3.2, we obtain that 2Δ varies from 0.011 to 0.088, that in absolute terms, is given as $2\hat{\Delta} = 2\Delta/(t^*_{cr} - t^*_0) = (0.32 - 2,56)^0C$.

Further, based on the formulation of the problem, the chosen solution origin along the coordinate x is limited by a cross-section x = ξ, where ξ - the unknown quantity, which is determined by the condition (3.20). Since for each possible numerical calculation, ξ is not previously defined, and there are no estimates of its value, then we must choose the interval of integration along x-axis that is certainly greater than the expected ξ. It is also possible that when our interval of integration would be smaller, we cannot get the information that we are interested in (i.e. the value of ξ).

Limitations of computer resources (speed and capacity of storage devices) may not allow to combine a large number of computational nodes along the y coordinate with taken with a stock number of nodes along the x coordinate. The above circumstances determine the following algorithm to solve this problem:

1. Neglecting the term $\frac{\partial}{\partial x}(\hat{\lambda}\frac{\partial \theta}{\partial \hat{x}})$ in equation (3.13), we reduce this equation to the parabolic form that allows using the effective implicit methods. In the numerical solution of a parabolic equation, in a finite-difference algorithm, the present values of

the unknown function are usually on two adjacent layers in parabolic coordinates (in this case, the coordinate $\hat{x}$). Therefore, we can make a decision on a dense grid, moving along the x coordinate from the cross section $\hat{x} = 0$ to a section in which $\theta_{out} < \theta_f$. This allows obtaining quickly and with a little expenditure of computer memory, an approximate temperature distribution in the layer of the product and, most importantly, the value of ξ.

2. The resulting temperature distribution is taken as the initial distribution for the solution of the full equation (3.13).

Thus, the finite-difference algorithm breaks down into a sequential solution of parabolic and elliptic equations.

We write the simplified equation of heat transfer neglecting conduction along the $\hat{x}$-axis direction, and a system of conditions of uniqueness (here and in what follows, all variables are dimensionless):

$$c_{eff}\frac{\partial \theta}{\partial x} = \frac{1}{Z_V}\frac{\partial}{\partial y}\lambda\frac{\partial \theta}{\partial y} \qquad (3.21)$$

Boundary conditions are:

$$(e_2/\lambda_{out})(\theta_{env} - \theta_{out}) = (\frac{\partial \theta}{\partial y})_{y=R+1} \qquad (3.22)$$

$$e_1\theta_{out}/\lambda_{pl} = (\frac{\partial \theta}{\partial y})_{y=0} \qquad (3.23)$$

Initial conditions are the following:

$$\theta_{x=0} = \begin{cases} \theta_{in}, R < y < R+1 \\ \theta_{env}(y/\lambda_{pl} + 1/e_1)/(1/e_1 + 1/e_2 + R/\lambda_{pl}), 0 \le y \le R \end{cases} \qquad (3.24)$$

We introduce a uniform finite-difference grid (Fig. 3.1):

$$(x_i, y_k), i = 1, 2, \ldots, \quad k = 1, 2, \ldots, M.$$

Let's approximate equation (3.21) by the implicit finite-difference scheme [49]:

$$c_{i+1,k}(\theta_{i+1,k} - \theta_{i,k})/\Delta x = [\lambda_{i+1,k+\frac{1}{2}} \cdot (\theta_{i+1,k+1} - \theta_{i+1,k}) -$$
$$\left.\begin{array}{c} -\lambda_{i+1,k-\frac{1}{2}}(\theta_{i+1,k} - \theta_{i+1,k-1})/\left[Z_v(\Delta y)^2\right] \\[4pt] 2 \le k \le M-1 \\[4pt] \lambda_{i+1,k+\frac{1}{2}} = (\lambda_{i+1,k+1} + \lambda_{i+1,k})/2 \\[4pt] \lambda_{i+1,k-\frac{1}{2}} = (\lambda_{i+1,k-1} + \lambda_{i+1,k})/2 \end{array}\right\} \quad (3.25)$$

Express the boundary conditions in the form of:

$$e_2\,(\theta_c - \theta_{i+1,M})/\lambda_{i+1,M} = (\theta_{i+1,M} - \theta_{i+1,M-1})/\Delta y; \quad (3.26)$$
$$e_1\theta_{i+1,1}/\lambda_{pl} = (\theta_{i+1,2} - \theta_{i+1,1})/\Delta y; \quad (3.27)$$
$$\theta_{1,k} = \begin{cases} \theta_{in}, R < y_k \le R+1 \\ \theta_{env}(y_k/\lambda_{pl} + 1/e_1)/(1/e_1 + 1/e_2 + R/\lambda_{pl}), 0 \le y_k \le R; \end{cases} \quad (3.28)$$

Net equation (3.25) approximates the differential equation (3.21) with an accuracy of 0 [Δx, $(\Delta y)^2$], and the equations (3.26) - (3.28) approximate, respectively, (3.22) - (3.24) with an accuracy of 0 (Δy).

To solve the system (3.25), we reduce it to the form

$$a_{i+1,k}\theta_{i+1,k+1} + \beta_{i+1,k}\theta_{i+1,k} + \gamma_{i+1,k}\theta_{i+1,k-1} = \delta_{i+1,k} \quad (3.29)$$

where

$$a_{i+1,k} = \lambda_{i+1,k+\frac{1}{2}}/\left[Z_v(\Delta y)^2\right]$$
$$\beta_{i+1,k} = -(\lambda_{i+1,k+\frac{1}{2}} + \lambda_{i+1,k-\frac{1}{2}})/\left[Z_v(\Delta y)^2\right] - c_{i+1,k}/\Delta x$$
$$\gamma_{i+1,k} = \lambda_{i+1,k-\frac{1}{2}}/\left[Z_v(\Delta y)^2\right]$$
$$\delta_{i+1,k} = -c_{i+1,k}\theta_{i,k}/\Delta x$$

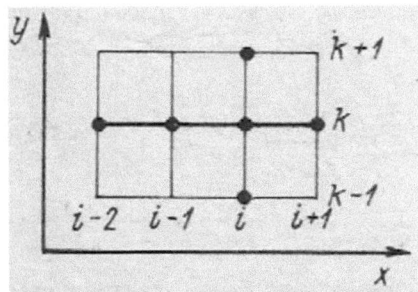

Fig.3.1 Computational grid for solving parabolic equation

System (3.29) can be conveniently solved by the sweep method [55]. For this purpose we represent the solution of $\theta_{i+1,k}$ in the form

$$\theta_{i+1,k} = A_{i+1,k}\theta_{i+1,k-1} + B_{i+1} \tag{3.30}$$

Substituting (3.30) into (3.29), we obtain the recurrence relations for the determination of the sweep coefficients

$$A_{i+1,k} = -\alpha_{i+1,k}/(\beta_{i+1,k} + A_{i+1,k-1}\gamma_{i+1,k}) \tag{3.31}$$

$$B_{i+1,k} = -(-\delta_{i+1,k} + \gamma_{i+1,k}B_{i+1,k-1})/(\beta_{i+1,k} + A_{i+1,k-1}\gamma_{i+1,k}) \tag{3.32}$$

From the boundary condition (3.27), this was transformed to the form

$$\theta_{i+1,1} = \theta_{i+1,2}/(1 + \Delta ye_1/\lambda_{pl})$$

by comparison with (3.30), we can obtain

$$A_{i+1,1} = 1/(1 + \Delta ye_1/\lambda_{cl}) \tag{3.33}$$

$$B_{i+1,1} = 0 \tag{3.34}$$

Thus, it is possible to find $A_{i+1,k}$ and $B_{i+1,k}$ for k = 2,3, ..., M-1; i.e. to make a direct moving of the sweep method.

In order to conduct reverse sweep method, it is necessary to determine $\theta_{i+1,\,M}$ from the boundary condition (3.26) and (3.30).

Indeed, on one hand, we have

$$\theta_{i+1,M} = \theta_{i+1,M-1}/(1 + \Delta ye_2/\lambda_{i+1,M}) + \theta_{env}\Delta ye_2/[\lambda_{i+1,M}(1 + \Delta ye_2/\lambda_{i+1,M})] \tag{3.35}$$

and, on the other hand –

$$\theta_{i+1,M-1} = A_{i+1,M-1}\theta_{i+1,M} + B_{i+1,M} \tag{3.36}$$

Combining the relations (3.35) and (3.36), we obtain an expression

$$\theta_{i+1,M} = \left(\frac{\frac{\Delta ye_2}{\lambda_{i+1,M}}\theta_{env}}{1 + \frac{\Delta ye_2}{\lambda_{i+1,M}}} + \frac{1}{\frac{\Delta ye_2}{\lambda_{i+1,M}}}B_{i+1,M-1}\right) * \left(1 - \frac{1}{1 + \frac{\Delta ye_2}{\lambda_{i+1,M}}}A_{i+1,M-1}\right) \tag{3.37}$$

Now you can, using the formula (3.30), make the reverse sweep method for k = M -1, M -2, ... , 2,1.

Having a strong diagonal dominance of the system of difference equations (3.29)

46

$$|a_{i+1,k}| + |\gamma_{i+1,k}| < |\beta_{i+1,k}| \qquad (3.38)$$

ensures the stability of the sweep method [55]:

The difference equation (3.29) is nonlinear with respect to $\theta_{i+1,k}$ because $c_{i+1,k}$, $\lambda_{i+1,k+1/2}$, $\lambda_{i+1,k-1/2}$ are depended on $\theta_{i+1,k}$:

$$c_{i+1,k} = \begin{cases} c_{pl}, \theta_{i+1,k} < 1 - \Delta, R < y_k < R+1; \\ r_v/(2\Delta) + (c_{pl}+1)/2, 1 - \Delta < \theta_{i+1,k} \leq 1 + \Delta, R < y_k \leq R+1 \\ 1, \theta_{i+1,k} > 1 + \Delta, R < y_k \leq R+1 \\ c_{pl}, \forall \theta_{i+1,k}, 0 \leq y_k \leq R \end{cases} \qquad (3.39)$$

$$\lambda_{i+1,k-\frac{1}{2}} = \begin{cases} \lambda_{sl}, \theta_{i+1,k+\frac{1}{2}} < 1; R < y_{k+\frac{1}{2}} \leq R+1 \\ 1, \theta_{i+1,k} \geq R, R < y_{k+\frac{1}{2}} \leq R \\ \lambda_{pl}, \forall \theta_{i+1,k+\frac{1}{2}}, 0 \leq y_{k+\frac{1}{2}} \leq R \end{cases} \qquad (3.40)$$

and for $\lambda_{i+1,k-1/2}$ - is similarly.

In accordance with this, the iterations are being carried out. Moreover, in order to find $\theta_{i+1,k}$ on f-iteration, the coefficients $c_{i+1,k}$, $\lambda_{i+1,k+1/2}$, $\lambda_{i+1,k-1/2}$ are calculated according to the values of $\theta_{i+1,k}$ on (f-1)-iteration.

For the numerical solution of equation (3.13), we introduce a uniform finite-difference grid:

$$(x_i, y_k), i = 1,2, \dots, IM; k = 1,2, \dots, KM$$

using a six-point scheme of second-order accuracy for x-axis and y (Fig. 3.2) [34]:

To express $c_{eff} (\partial\theta/\partial x)$ we use an "upstream"approximation:

$$c_{eff}(\partial\theta/\partial x) \approx [2 \cdot (\theta_{i,k} - \theta_{i-1,k})/(\Delta x) - (\theta_{i,k} - \theta_{i-2,k})/(2 \cdot \Delta x)] \cdot c_{i,k} + 0[(\Delta x)^2] \quad (3.41)$$

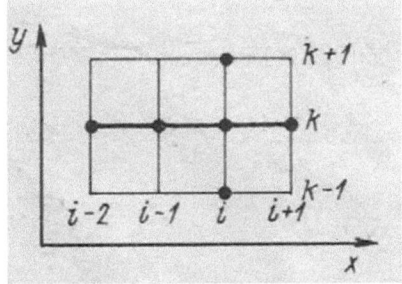

Fig.3.2 Computational grid for solving elliptic equation

We approximate the right-hand side of equation (3.13) by a symmetrical pattern of the second order of accuracy

$$\frac{1}{Z_v}\left(\frac{\partial}{\partial x}\lambda\frac{\partial\theta}{\partial x} + \frac{\partial}{\partial y}\lambda\frac{\partial\theta}{\partial y}\right)$$

$$= \frac{1}{Z_v}\left\{\frac{1}{(\Delta x)^2}\left[\lambda_{i+\frac{1}{2},k}(\theta_{i+1,k} - \theta_{i,k}) - \lambda_{i-\frac{1}{2},k}(\theta_{i,k} - \theta_{i-1,k})\right]\right.$$

$$\left. + \frac{1}{(\Delta y)^2}\left[\lambda_{i,k+\frac{1}{2}}(\theta_{i,k+1} - \theta_{i,k}) - \lambda_{i,k-\frac{1}{2}}(\theta_{i,k} - \theta_{i,k-1})\right] + 0[(\Delta x)^2,(\Delta y)^2]\right\} \quad (3.42)$$

Equating equations (3.41) and (3.42), we obtain a finite-difference equation for determining $\theta_{i,k}$. To solve this equation we will use the relaxation method [44, 56, 70]. We introduce an iterative index j.

Let's organize the following computer iterative process

$$\theta_{i,k} = (\pi_{i,k} - \beta_{i,k}\theta^j_{i+1,k} - \gamma_{i,k}\theta^j_{i-1,k} - \delta_{i,k}\theta^j_{i,k+1} - \varepsilon_{i,k}\theta^j_{i,k-1})/\alpha_{i,k} \quad (3.43)$$

$$\theta^{j+1}_{i,k} = \theta^j_{i,k} + \eta(\theta_{i,k} - \theta^j_{i,k}) \quad (3.44)$$

where

$$a_{i,k} = -[(\lambda_{i+\frac{1}{2},k} + \lambda_{i-\frac{1}{2},k})/(\Delta x)^2 + (\lambda_{i,k+\frac{1}{2}} + \lambda_{i,k-\frac{1}{2}})/(\Delta y)^2]/Z_v - 2c_{i,k}/\Delta x \quad (3.45)$$

$$\beta_{i,k} = \lambda_{i+\frac{1}{2},k}/[Z_v(\Delta x)^2] \quad (3.46)$$

$$\gamma_{i,k} = \lambda_{i-\frac{1}{2},k}/[Z_v(\Delta x)^2] + 2c_{i,k}/\Delta x \quad (3.47)$$

$$\delta_{i,k} = \lambda_{i,k+\frac{1}{2}}/[Z_v(\Delta x)^2] \quad (3.48)$$

$$\varepsilon_{i,k} = \lambda_{i,k-\frac{1}{2}}/[Z_v(\Delta y)^2] \quad (3.49)$$

$$\pi_{i,k} = -c_{i,k}(\theta^{j-1}_{i,k} - \theta^{j-1}_{i-2,k})/(2\Delta x) \quad (3.50)$$

for all interior points of the field $3 \leq i \leq IM -1$, $2 \leq m \leq KM -1$.

Here, η - is the coefficient of relaxation. Preliminary numerical calculations have shown that $\eta = 0.5$ ensures the convergence of the iterative process for the valid values of the obtained similarity criteria and dimensionless conversion factors.

After calculating $\theta^{j-1}_{i,k}$ at all interior points, there are defined the values of $\theta^{j+1}_{i,k}$ on the boundaries, based on the finite-difference approximation of the boundary conditions:

$$e_2(\theta_{env} - \theta_{i+1,KM})/\lambda_{i+1,KM} = (\theta_{i+1,KM} - \theta_{i+1,KM-1})/\Delta y \quad (3.51)$$

$$e_1\theta_{i+1,1}/\lambda_{pl} = (\theta_{i+1,2} - \theta_{i+1,1})/\Delta y \qquad (3.52)$$

$$\theta_{1,k} = \begin{cases} \theta_{in}, R < y_k \le R+1 \\ \theta_{env}(y_k/\lambda_{pl} + 1/e_1)/(1/e_1 + 1/e_2 + R/\lambda_{pl}), 0 \le y_k \le R \end{cases} \qquad (3.53)$$

$$(\theta_{IM,k} - \theta_{IM-1,k})/\Delta x = 0 \qquad (3.54)$$

Condition for the stability of the computational process is the inequality [25]

$$\left|\beta_{i,k}\right| + \left|\gamma_{i,k}\right| + \left|\delta_{i,k}\right| + \left|\varepsilon_{i,k}\right| \le \alpha_{i,k} \qquad (3.55)$$

Substituting (3.45) - (3.49) in (3.55), we see that this inequality is satisfied.

Flowchart of the numerical investigation of temperature fields of the interacting bodies (product - a wall) is shown in Fig. 3.3.

As an example, Fig. 3.4 shows a fragment of the temperature field of a composite body "product-wall" with "average" values of the parameters of the process.

Realized in the form of a computer program, the mathematical model (in particular, model of the thin material freezing on a moving cooled wall) is a kind of computational experimental unit [56] that has several advantages over conventional technology experimental construction:

- universality, because for the study of a new version by the computing installation there is only necessary to introduce a new background information, while at the technically realized experiment it will need a lot of raw materials, and sometimes reinstalling, reconstruction and even full-scale installation of the new design;

- possibility to obtain complete information about the effect of process parameters on

the temperature field of the interacting bodies.

However, an array of information provided by the computing unit has a very large volume, making it difficult to process it.

At the same time, implementation of full-scale experiment at the test conditions of the process equipment would be fraught with even greater difficulties. In order to be able to compare the numerical calculations and the experimental data, it is necessary to hold at least the same number of experiments, variants as calculated by the computing unit. To make the experimental data statistically significant, it is needed to organize in each experiment, 3-5 replications. This will lead to an even larger increase in labor costs, increase in terms of the experiments, which, in turn, affect the accuracy of the experimental data.

On the other hand, it is obvious that at random, haphazard use of any sorting options, usage of the fastest computer does not provide optimal solutions. It needs the deliberated and planned considing of these options.

Recall that by using the criteria of similarity and dimensionless transformation factors (Table. 3.2), the solution of the system of equations (3.13) - (3.20) can be written as

$$\xi = \Phi\left(z_v, \hat{r}_{v,} e_1, e_{p_2}, \theta_{in}, \theta_{env}, \theta_f, \widehat{R}, \hat{c}_{sl}, \hat{c}_{pl}, \hat{\lambda}_{sl}, \hat{\lambda}_{pl}\right) \tag{3.56}$$

moreover, a particular form of the function Φ is not known beforehand.

Hence it is clear that the objective function ξ depends on a large number of parameters and the solution of the problem is a solving of a multi-task. It is obvious that its decision is extremely difficult.

However, not all parameters are equally affecting the researched process. So the reduction of the number of parameters to a minimum on the basis of their relative influence and selection of essential process parameters is the most important goal in the correct formulation of the problem. For this reason the active principles of the theory of experimental design [57] are most valuable, specially, the method of random balance.

In the method of random balance, linear effects and pair wise interactions are eliminated. But, at the same time, there is an additional constraint: it is assumed that the number of significant effects is significantly less than the total number of effects taken under suspicion.

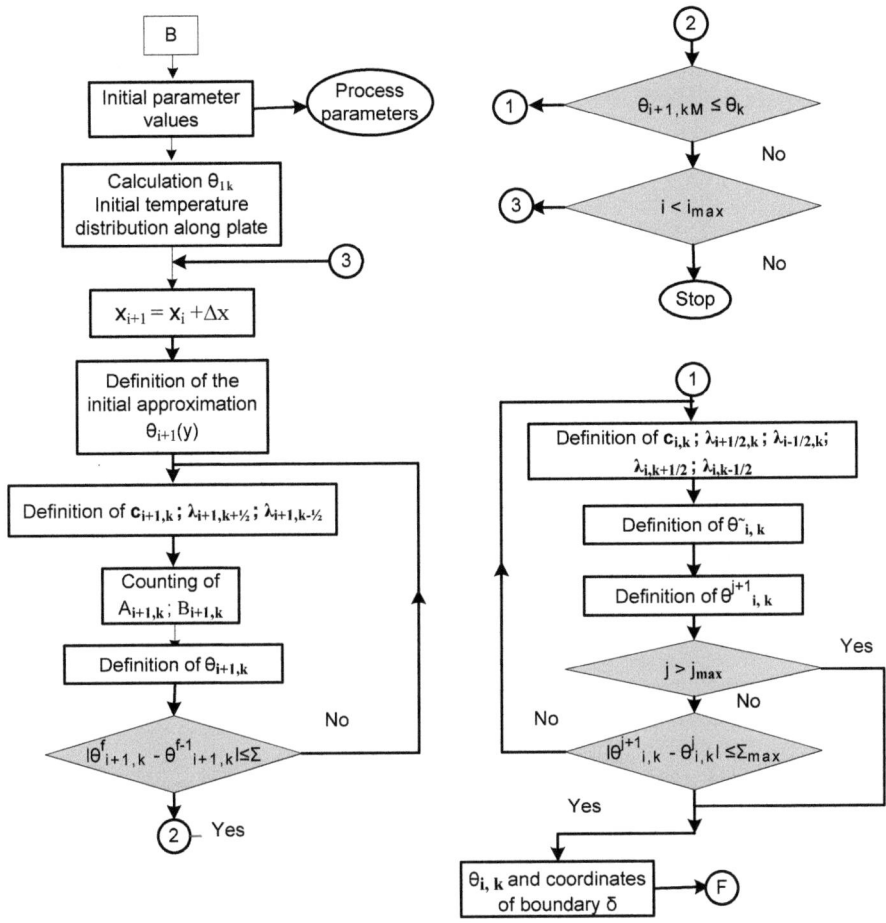

Fig.3.3 Flowchart of the numerical investigation of temperature fields
of the interacting bodies (product - a wall)

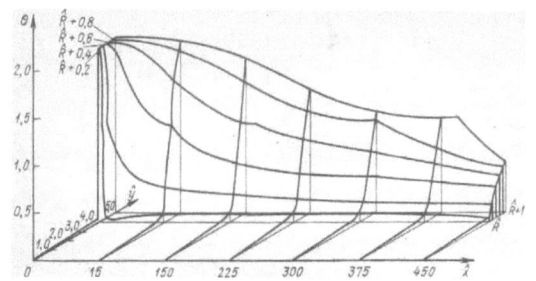

Fig.3.4 Detail of the temperature field of the interacting bodies
at "zero" values of the parameters of the process

Application of random balance in the study of the freezing process of a thin material layer on the cooled wall (as well as any other process) has, in principle, two features.

The solution to any practical problem will be of great value when the independent variables are used as generalized criteria, rather than individual factors of the physical dimension. The rationale for this approach is justified in [58]. In this case, the monitoring process is less sensitive to variations of similarity criteria, rather than to the combined effect of variations of the parameters in the similarity criteria. Application of the theory of similarity to solve problems by using the theory of experiment planning is due to the desire to reduce the number of independent variables, and, therefore, number of experiments, and dramatically reduce the amount of computational work.

Another feature is the fact that all methods of the theory of experimental design, including the method of random balance, are used in a full-scale natural experiment. In this paper we use the method of random balance to identify significant factors in the framework of the developed mathematical model. Such approach from the standpoint of mathematics does not have currently theoretical studies. However, from an engineering point of view, by the condition of the availability of positive experimental evaluation, which is, of course, only a partial justification the suggested approach to the problem of freezing of continuous product layer on a cooled wall, is seemed a possible.

Based on the above approach to the development of PMM, we can formulate the tasks that must be solved during the computational experiment: to identify the relative influence of operating and design parameters/variables of DF on the solidification process and sub-cooling of the product layer to the desired final temperature; to obtain analytical relationships that take into account the essential parameters of the test process and leading to the development of a rational design of DF.

3.2. Calculation of freezing and sub-cooling of layer of pastor and minced-shaped product

In order to reduce the amount of computational work on the computer, to facilitate the processing of the results by reducing the number of influencing parameters on the basis of their relative influence and identification of the essential parameters of the test process, we use the method of random balance.

In the method of random balance [59] there are used supersaturated plans in which the number of trials (experiments) μ is less than the number of the effects, but is higher than the amount of significant effects q ($\mu > q$).

This method is used to determine the most significant factors that characterize the object under study [60].

The applying the method of random balance is based on two assumptions: 1) if for the development of the experimental plan, one uses random sampling of the rows of full factorial experiment, then this random mixing, the probability of separating of dominant effects will be great enough just because of the small number of these effects; 2) factors do not affect on the response of the system, i.e. they can be ranked (ranking is exponential) in descending order of influence on ξ, and while most of them can be attributed to the noise background.

Compliance of the condition μ-q ≥ 0 gives a possibility of the quantity measure of the chosen effects by regression analysis [60].

The abundance of the most detailed information obtained from numerical studies on the basis of the developed finite-difference algorithm (see chapter 2.3) is not always necessary, or rather, almost never required to produce correct and effective design solutions. In each case, it is necessary to understand clearly for what purpose should hold one or the other option for the computing calculation. Detailed information is sometimes just as harmful as it is not permissible to-consolidation and generalization. Therefore, in the development of design solutions, it is very important a skill to carry out the required level and in the right scale aggregation of information (i.e., its consolidation in order to reduce the amount of information considered).

From this perspective, the use of the random balance will highlight the significant factors, the number of which is sufficient for the desired amount of design studies in the shortest possible period of time. The boundaries of the numerical investigation are determined by the value of the similarity criteria and dimensionless transformation factors (see Table 3.2).

The procedure for processing the results of screening experiments is perceived difficulty just by reading its description. In order to be able for further use of the above described technique, the first step of sifting of significant factors will be explained in detail[2].

To construct the plan of experiments, we use the mixing of random samples of fractional factorial designs.

We encode the similarity criteria and dimensionless transformation factors (Table. 3.3).

We divide the factors into the following groups: 1) X_1 X_2 X_3 X_4; 2) X_5 X_6 X_7 X_8; 3) X_9 X_{10} X_{11} $X1_2$. Further we use the type of experiment plans 2^4. By usage of random numbers from the plan table 2^4 for each group of factors, we randomly choose a line (16 experiments - 16 lines) [57]: *Group 1*: 10.6 14.1 3.16 4.15 5.9 7.13 2.8 12.11; *Group 2*: 13.3 7.16 2.9 2.16 6.12. 13.7 9.12 3.6; *Group 3*: 15,16 5,10 15,4 3,4 6,9 9,3 16,5 6,10. We form a matrix design of experiments (Table. 3.4). It defines the conditions and sequence of numerical calculations.

Table 3.3. Coding of similarity criteria and dimensionless conversion factors

X_1	X_2	X_3	X_4	X_5	X_6	X_7	X_8	X_9	X_{10}	X_{11}	X_{12}
Z_v	θ_{in}	$\hat{R}$	e_1	$\hat{r}_v$	θ_{env}	e_{p2}	$\hat{c}_{sl}$	$\hat{c}_{pl}$	$\hat{\lambda}_{sl}$	$\hat{\lambda}_{pl}$	θ_f

Scatter diagram of the results of three phases of screening experiments are shown in Fig. 3.5. The correctness of the diagram constructing is checked by counting the dots on the right and left of each of the perpendicular - each side must be with eight points, as each column of the matrix contains eight minus signs and eight plus signs. Provisions of medians [62] are found as the arithmetic mean between the 4th and 5th points of each histogram. Select points are marked by circles; their number for each factor is given by the corresponding perpendicular.

---

[2] Algorithms machining screening experiments described, for example, in [61]

Table 3.4. Matrix of the experiment design

#	X_1	X_2	X_3	X_4	X_5	X_6	X_7	X_8	X_9	X_{10}	X_{11}	X_{12}
1	+	-	-	+	-	-	+	+	-	-	+	-
2	+	-	+	-	-	+	-	-	+	-	-	+
3	-	+	-	-	-	+	+	-	-	+	+	+
4	+	+	+	+	+	+	+	+	+	+	+	+
5	+	+	-	-	+	-	-	-	-	+	-	-
6	-	+	+	+	-	-	-	+	+	+	-	-
7	+	-	+	+	+	-	-	-	-	+	+	+
8	-	-	-	-	+	+	+	+	+	+	-	-
9	-	-	+	-	+	-	+	-	+	-	+	-
10	-	-	-	+	+	+	-	+	-	-	-	+
11	-	+	+	-	-	-	+	+	-	-	-	+
12	-	-	+	+	-	+	+	-	-	+	-	-
13	+	-	-	-	-	-	-	+	+	+	+	+
14	+	+	+	-	+	+	-	+	-	-	+	-
15	+	+	-	+	-	+	-	-	+	-	+	-
16	-	+	-	+	+	-	+	-	+	-	-	+

Table 3.5 shows the results and the successive stages of their adjustment as "removing" significant effects. The value of ξ, corresponding to the number of experience u, is denoted by $\mathbf{y_u}$.

Table 3.5. The results of screening experiments and the successive stages of their adjustment

#	Ist stage		IInd stage			IIId stage			
u	y_u	Δy_u^I	y_u^I	y_u^I+120	Δy_u^{II}	y_u^{II}	$y_u^{II}+320$	Δy_u^{III}	y_u^{III}
1	756	874.25	-118.25	1.75	0	1.75	321.75	-301.8	623.55
2	556	398.5	167.50	277.50	0	277.5	597.50	-37.0	967.55
3	156	-475.75	631.75	751.75	500	251.75	571.75	-301.7	373.55
4	886	398.5	487.50	607.50	924.5	-317.00	3	-671.8	674.80
5	2,770	874.25	1,895.75	2,015.75	924.5	1,091.25	1,411.25	0	1,411.25
6	184	0	184.00	304.00	500	-196.00	124.00	-370.0	494.00
7	618	398.5	219.50	339.50	424.5	-85.00	235.00	-301.8	536.80
8	212	0	212.50	332.00	424.5	-92.50	227.50	-370.0	597.50
9	218	0	218.00	338.00	424.5	-86.50	233.50	-671.8	905.30
10	216	-475.75	691.75	811.75	424.5	387.25	707.25	0	707.25
11	224	-475.76	699.75	819.75	500	319.75	639.75	0	639.75
12	208	0	208.00	328.00	0	328.00	648.00	0	648.00
13	376	398.5	-22.50	97.50	0	97.50	417.50	-671.8	1,089.30
14	1,692	874.25	817.75	937.75	924.5	13.25	333.25	-301.8	635.05
15	1,006	874.25	125.75	245.75	500	-254.25	65.75	-671.8	737.55
16	248	-475.75	723.75	843.75	924.5	-80.75	239.25	-370.0	609.25
$S^2\{y_u\}$	993,288		323,341			117,641			56,740
$S^2\{\bar{y}_u\}$	997		482			343			238

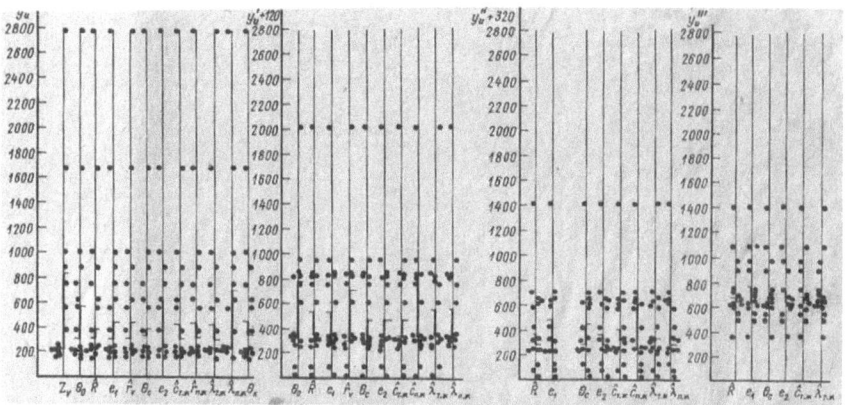

Fig.3.5 Scatter diagram of the results of three phases of screening experiments

As seen from Fig. 3.5a, maximum displacement of the medians is found for factors X_1, X_7, X_{11}, and the largest displacement of histograms, determined by the number of points stand out – corresponds to factors X_1, X_5, X_7, X_{12}. For the final selection we use the arithmetic mean criterion [59]:

$\bar{B}_1$	$\bar{B}_5$	$\bar{B}_7$	$\bar{B}_{11}$	$\bar{B}_{12}$
3,976	1,680	1,214	1,386	2,008

$\bar{B}_1$ and $\bar{B}_{12}$ have the highest values. We organize a statistical analysis of the factors X_1 and X_2 (Table. 3.6).

The regression coefficient for the selected factors is calculated by the formula

$$b_1 = (\sum_{1}^{c} x_i^R \, y^{-R})/C; \; y^{-R} = \sum_{1}^{l} y_u^R /l$$

We know that $l = 4$, $C = 4$, then

$$b_1 = (-822/4 + 6224/4 - 844/4 + 2436/4)/4 = 437,125$$
$$b_{12} = (-822/4 - 6224/4 + 844/4 + 2436/4)/4 = -237,875$$

To calculate the variance $S^2\{b_i\}$ there are used the following values

$$\sum y_u^l \sum (y_u^l)^2, \varepsilon^2 = \sum (y_u^R)^2 - (\sum y_u^R)^2/l \quad \text{(look Table 3.6)}$$

The dispersion of single value

$$S^2\{y\} = (\sum_{1}^{c} \varepsilon^2)/[\sum_{1}^{c} (l-1)] = 214,496.2$$

The residual dispersion

$$S_p^2 = S^2\{y\} \sum_{1}^{c} (1/l) = 214,496.2$$

The dispersion of the regression coefficients

$$S^2\{b_i\} = S_p^2/C^2 = 13,406.1$$

Student test for the number of degrees of freedom

$$f_t = \sum_{1}^{c} 1 - C = N - C = 12$$

The magnitude of t-criteria at the 5% significance level is 2.18 [62]. Given the low sensitivity of the method of random balance, usually in the calculations there is chosen the value of t-criteria equaled 2, i.e.

$$|b_i| > t\sqrt{S^2\{b_i\}} \approx 2\sqrt{S^2\{b_i\}}$$

Table 3.6. The results of the first stage of screening experiment

Matrix X^I full factorial experiment 2^2		Number of experiments in matrix	Matrix of results y_u	$\Sigma\, y_u$	$\Sigma\, (y_u)^2$	ε^2	??? y_u	$\Delta\bar{y}$
X_1	X_{12}							
-	-	6; 8; 9; 12	184; 212; 218; 208;	822	169,588	667	205.5	0
+	-	1; 5;14; 15;	756; 2,770; 1,682; 1,006;	6,224	1,211,9330	2,434,702	1,556	874.,25
-	+	3; 10; 11;16;	156; 216; 248; 224;	844	182,672	4,558	211	475.75
+	+	2; 4; 7; 13;	556; 886; 618; 376;	2,436	1,617,432	133,908	609	398.5

In this case, if $S\{b_i\} = 115,78$, the statistically significant regression coefficients are all greater than 231.56. Consequently, $b_1 = 437.125$ and $b_{12} = -237.875$ are recognized as the significant.

Exclude the effect of X_1 and X_{12}. To do this, we calculate

$$\Delta y_u = b_1 \cdot (x_{1u} + 1) + b_2 \cdot (x_{2u} + 1) + \ldots + b_k \cdot (x_K + 1)$$

where k - total number of extracted factors, i.e.,

$$\Delta y^I = b_1 \cdot (x_{1u} + 1) + b_2 \cdot (x_{2u} + 1)$$

The calculation results are given in Table. 3.6, the corrected values $y_u^I = y_u$ $-\Delta y_u^I$ are in Table 3.5.

Correct selections of factors X_1 and X_2 must be confirmed by a decrease in the scattering results of y_u^I. The dispersion of the results of numerical experiments equals

$$S^2\{y_u\} = \left[\sum y_u{}^2 - \left(\sum y_u\right)^2/N\right](N-1)$$

$$S^2\{y_u\} = (15,619,020 - 11,515,070/16)/(16-1) = 993,288.5$$

$$S^2\{y_u\} = 996.64$$

Then the variance of the results after the first stage separation of significant factors will be

$$S^2\{y_u^I\} = 232,341.1; S\{y_u^I\} = 482.02$$

According to the results $y_u^I + 120$, there is built a scatter diagram shown in Fig. 3.5b. The subsequent stages (second and third) are similar to the first stage of selection. The scatter plots of the results of the second and third phases are shown in Fig. 3.5, c, d.

As a result of screening experiments in three consecutive rounds, six essential factors are identified. They are the process parameters of a thin freezing of pastor and minced-shaped products on a moving cooled wall: Z_v - a criterion that takes into account the motion of the wall with respect to the material coordinate system that rigidly connected with the place of the applying the product on a cooled surface; $\hat{r}_v$ - a criterion that takes into account the latent heat of phase changing; θ_f - final temperature on the outside (in contact with the environment) surface of the product; θ_{in} - initial temperature of the product being a cold treatment; $\hat{c}_{pl}$ - volumetric heat capacity of the cooling wall; $\hat{\lambda}_{pl}$ thermal conductivity of the cooling wall.

Here it is appropriate to make a few observations on the further procedure of processing the results of the numerical experiment.

The random balance is the first step in the experiment. Subsequently, there are performed "movement in the area and a description of the area of the optimum" [57].

Time of termination of the sifting of effects in the random balance method is assessed by using Fisher's criterion [62]

$$F = S^2\{y_u\}/S^2\{y\} \quad (3.57)$$

where $S^2\{y_u\}$ - the dispersion of the results of experience; $S^2\{y\}$ – variance calculated based on the results of several parallel experiments in the center of the experiment ("reproducibility variance").

Sieving effects is cease if the value F calculated by the formula (3.57) is less than the table value for the selected level of significance, i.e., it founds that the remaining variation of points is not different from the scattering results related to the experimental error.

In view of the fact that the computer experiment is conducted on models of different physical nature than the thermal process under study [63], the notion of "reproducibility variance" appears incompetent.

It should be noted that, despite the positive results of applying the methods of the experiment planning in the creation of mathematical models of complex physical objects, there is a problem to reduce the time and cost for the required number of experiments in the subsequent stages of finding of the optimum area.

The complexity of solving this problem is determined primarily by the inability of the a priori definition of the order of the mathematical model due to the complexity of the physical nature of processes [64], as well as the fact that for every freezable material there is its own region of the optimum. In addition, because of the essential nonlinearty of functions that linking ξ with parameters of the investigated process, and high accuracy requirements to describe the area of the optimum, it is required to build mathematical models of higher orders. This in turn makes it difficult, and sometimes makes it impractical to use the obtained results in practice.

To this we must add that the qualitative and quantitative set of physical parameters, including factors (the variables of the process under study) selected based on the results of three phases of the sieve numerical experiment, is sufficient for engineering design calculations of drum machines to determine the achievable performance of the designed device for each type of material.

Given the above, the finding of the analytic dependence of the main function ξ from the significant effects is realized on the results of the third stage of separating the essential factors of the investigated process.

Knowledge of the temperature distribution in the layer of the product, the nature of change in the position of the phase change boundary $\sigma(\sigma = \hat{y} - R, \hat{R} < \hat{y} \le \hat{R} + 1)$, and the length of the zone in which the temperature of the outer surface of the product is super cooled, allow to give an indication of the qualitative and quantitative impact of selected parameters on the process of freezing of the material thin layer on a cooled wall. It is eventually a necessary condition for designing of the drum-type freezer.

Consider the effect of initially selected parameters (Z_v, θ_f, θ_{in}, $\hat{r}_v$, $\hat{c}_{pl}$, $\hat{\lambda}_{pl}$)) on the temperature of the external surface of the layer θ_{out} (Fig. 3.6), exploring the

60

nature of the change along the x-axis. All graphs can be divided into three zones: a first zone corresponds to the cooling step of the material, the second zone - the stage of the phase change when $\theta_{out} = 1$, and the third zone - the step of super cooling of the product to a predetermined temperature. In the allowable ranges of variation of parameters (Table. 3.2) at distances x<20, θ_{out} practically remains constant for all test modes of DF. The more criteria Z_v (Fig. 4.6,a), the greater the distance from the place of applying a layer of the product, on which the condition $\theta_{out} = 1$ is achieved. With increasing Z_v, θ_{out} is reduced in the first zone according to the law that is close to linear. In the third zone (frozen material, $\theta_{out} < 1$) with $Z_v < 300$, the temperature of the layer outer surface varies linearly. At the same time, when $Z_v > 300$ the heat transfer increases along the x-axis, and the distribution along the x acquires hyperbolic.

θ_{out} distribution along the x-axis, depending on θ_{in} (Fig. 3.6,b) and $\hat{r}_v$ (Fig. 3.6c) is described similarly. However, these parameters are less than Z_v, influence on the length of the cooling, freezing and sub-cooling zones of the product.

The change of θ_f does not affect on θ_{out} (see Fig. 3.6,d).

In contrast, the increase of $\hat{c}_{pl}$ (Fig. 3.6,e) or $\hat{\lambda}_{pl}$ (Fig. 3.6,f) reduces the freezing and sub-cooling zone of the product layer ξ. However, by changing these parameters within acceptable ranges of variation, the influence is very weak.

These results correspond to well-known theoretical ideas about the progress of physical processes, accompanied by absorption (release) of heat [65].

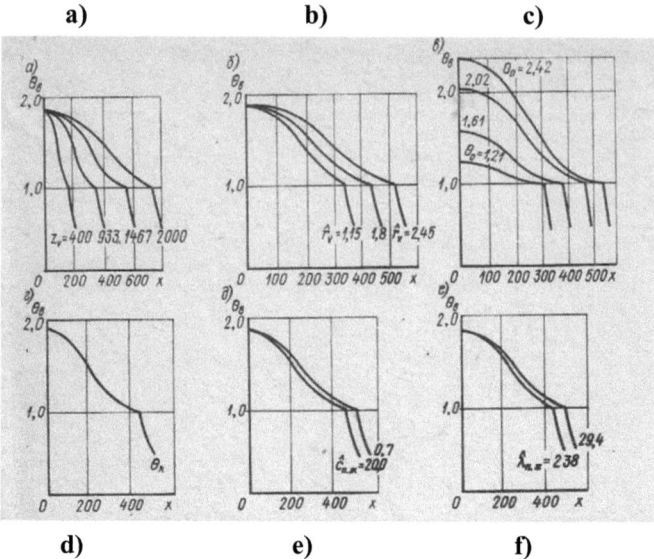

Fig.3.6 Temperature distribution of outer surface of the product layer along x-axis, depended on the selected parameters: a - Zv, b- $\hat{r}_v$, c - θout, d - θf; e - $\hat{c}_{pl}$, f- $\hat{\lambda}_{pl}$

Analysing the dependence of the front position of the phase transfer σ along x from the selected parameters allows finding the inherent following general patterns. For x<40 there is a "disordered" regime, characterized by a sharp impact of the initial state of the system of interacting bodies (product - a wall) on its temperature field. In the thickness of the material there are significant temperature gradients. The position rate of the phase change front is A = dσ/(dx) -> ∞.

At a distance x> 40 the influence of the initial temperature distribution in the product and the wall for their further change is smoothed that corresponds to the case A=dσ/(dx)=const. Certainly, the value of A is determined by the thermal characteristics of the interacting bodies, their size and the conditions of heat transfer.

The degree of influence of the significant parameters on σ(x) is consistent with previous estimates. As already indicated, within the proposed PMM due to results of screening experiments, six essential factors of the process of the material layer freezing on a cooled wall with continuous renewal of the initial conditions were identified. Simplified criterion equation can be written as follows:

$$\xi = \Phi^*(Z_v, \theta_f, \theta_{in}, \hat{r}_v, \hat{c}_{pl}, \hat{\lambda}_{pl})$$

For a more detailed study of the researched system, the target function f can be represented in the form of the correlation function of the one-parameter chosen functions:

$$\xi = K' \cdot f_1(Z_v) \cdot f_2(\theta_f) \cdot f_3(\theta_{in}) \cdot f_4(\hat{r}_v) \cdot f_5(\hat{c}_{pl}) \cdot f_6(\hat{\lambda}_{pl}) \quad (3.58)$$

Finding a specific type of function f_s (s = 1,2, ..., 6) and a constant factor K' is organized by varying each individual parameter in the valid range of its values and the followed treatment of the numerical results obtained by the method of least squares, while maintaining other essential parameters at the average level. The remaining six non-essential factors are fixed at the average level.

We obtained the following relationship:

$$f_1(Z_v) = 0{,}39 \cdot (Z_v + 25{,}64) \quad (3.59)$$

$$f_2(\theta_f) = 437.03 \cdot \theta_f^{-0.181} \quad (3.60)$$

$$f_3\theta_{in} = 153{,}09 \cdot \theta_{in} + 191.8 \quad (3.61)$$

$$f_4(\hat{r}_v) = 143{,}02 \cdot \hat{r}_v + 231.22 \quad (3.62)$$

$$f_5(\hat{c}_{pl}) = 541 \cdot \hat{c}_{pl}^{-0.05} \quad (3.63)$$

$$f_6(\hat{\lambda}_{pl}) = 617 \cdot \hat{\lambda}_{pl}^{-0.051} \quad (3.64)$$

$$K' = 3.98 \cdot 10^{-14} \quad (3.65)$$

Substituting (3.59) - (3.65) in (3.58), by simple transformation the criterion equation (3.58) reduces to

$$\xi = 0{,}05 \cdot (Z_v + 25{,}64) \cdot \theta_f^{-0.181} \cdot (\theta_{in} + 1{,}253) \cdot (\hat{r}_v + 1{,}617) \cdot \hat{c}_{pl}^{-0.05} \cdot \hat{\lambda}_{pl}^{-0.051} \quad (3.66)$$

Formula (3.66) gives a clear picture of the qualitative and quantitative impact of significant factors on the freezing and sub-cooling processes of the product layer on the cooling surface. Let's more detail the possible explanations for elimination of the unimportant factors.

The insignificance of complex e_1 can be explained by the fact that at the reached value of film coefficient from the refrigerant α_o (over 1,000 W/(m²·K)), the thermal resistance of contact "refrigerant-to-wall surface" can be neglected in comparison with the thermal resistance of the layer of unfrozen product.

Little impact of complex e_2 can be explained by the following. At the implemented in practice working modes, the coefficient α_{env} is typically 5-20

W/(m²·K), and hence, the thermal resistance of contact "product layer - the environment" is many times greater than the thermal resistance of the material layer. That is why, the heat transfer with the environment, at the sufficient accuracy for engineering calculations, can be neglected. This also explains the insignificance of θ_{env}.

Allocation of volume heat capacity of the frozen material c_{sl} to insignificant parameters is explained by more heat-accumulating capacity of unfrozen material compared to the solidified.

The insignificance of the dimensionless coefficient of thermal conductivity of the frozen layer of material $\hat{\lambda}_{sl} = \lambda_s/\lambda_l = (h/\lambda_l)/(h/\lambda_s)$ can be explained by the smallness of the thermal resistance of the frozen layer as compared with unfrozen.

The thickness of the drum wall $\hat{R}$ affects the intensity of heat transfer and freezing of the thin-film materials. However, within the range of commonly accepted in practice in order to ensure the strength of the drum-evaporator, the effect of R on the extent of the zone ξ is immaterial, as it is shown by the analysis. Implicitly, the thickness of the wall is still present in the final analytical formula (3.66): $\hat{R}$ is included in the complex $\hat{\lambda}_{pl} = (h/\lambda_l) \cdot (R/h)/(R/\lambda_p)$, which characterizes the ratio of the thermal resistance of the layer of the product and the wall at a predetermined relative thickness of the drum-evaporator wall.

Thus, the results of numerical studies suggest the following conclusions.

1. When calculating the ξ, in the framework of the proposed PMM, it is necessary to take into account the change in the permissible ranges of variation of only part of the variables that affect on the freezing and super cooling process of the thin layer product to the desired final temperature on the moving cooled wall.

2. The proposed solution that derived from a mathematical model based on general physical laws does not contain any empirical coefficients. It is a rational[3] model for the process calculation of thin-layer freezing when Z_v takes large values ($Z_v \gg 1$).

3. Change of the wall thickness at the specified range of variation, in spite of the current point of view, has a little effect on the process under study. Therefore, the choice of the wall thickness of the drum evaporator should be made only in accordance with the existing regulations of the safety of refrigeration systems or, for example, Lloyd Register.

---

[3] It does not address issues related to the proof of the existence, uniqueness and stability of the optimal solution. Therefore, actually, the contents of chapter correspond to the search of non-optimal (in the strict sense), but, if possible, rational modes of the material layer freezing on a cooled wall.

Performed experimental reseaches [66] confirmed the acceptability of the studying of:

a. the processes of thin-layer freezing of materials with different thermal characteristics used in the mathematical model;

b. the developed finite-difference algorithm and the methods of the theory of active planning of experiments at the processing of the numerical results obtaining.

3.3 Experimental studies of thin-layer freezing product on a moving cooled wall

The experimental studies are based on the principle of closest approach of the laboratory experiment to the industrial conditions of DF. When conducting a physical experiment, the dimensional variable l - length of the zone of freezing and super cooling of the material was constant. There were varied the thickness of the product layer h and the speed n of the drum-evaporator. This is because, in practice, the final temperature of the outside surface of layer t_f can be measured with greater precision than l. It is a constant: length of "zone" is restricted by the place of the layer supplying on the surface of the drum and the location of the shear device, by which the frozen layer product is removed from the drum surface. In addition, the requirements for the quality of food technologists, in a certain way, are also associated with a final temperature [67]. Therefore, all experiments were arranged in a sequence that provides the ability to compare the temperature measured in experiments with a final temperature calculated from the theoretical formula (3.66):

$$* [\frac{0.05(\frac{\pi Dhnc_l}{\lambda_l} + 25.64)(\frac{t_{in} - t_0}{t_{cr} - t_0} + 1.253)(\frac{r_v W \omega}{c_l(t_{cr} - t_0)} + 1.617)}{\frac{\pi Dp}{h^*}(\frac{c_p}{c_l})^{0.05}(\frac{\lambda_p}{\lambda_l})^{0.05}}]^{5.25} \quad (3.67)$$

where $p = l/(\pi D)$ - coefficient of the cooling surface utilization   that takes into account the actual proportion of work surface with respect to the total surface of the drum-evaporator.

The usage of this factor is due to several reasons. It is known that at the development of any technical systems, designers strive to achieve the perfect end result, in this case, the use of the full entire working surface of the drum-evaporator. Unfortunately, in practice, this is not feasible, since: a) some of the surface of the drum is busy by means, placed over it, for applying a layer of material and removing it after freezing; b) at working conditions of drum freezers (especially on ships) one cannot ensure the stability of the temperature regimes of the device, the constancy of the thermo-physical properties of the processed materials.

Possible range of p variation can be found from the condition $p = 1 - \delta\xi$, where $\delta\xi$ – the relative maximum extent of the zone of freezing and sub-cooling of the product layer. To calculate $\delta\xi$, there was used a formula given in [68].

Calculations have shown [66] that the value of the coefficient p is in the range of 0.51 - 0.75, the upper limit is almost unreachable because corresponds to the "accurate" knowledge of the thermophysical characteristics included in the essential parameters of the investigated process.

For a visual comparison of the results of theoretical calculations and experiments, the theoretical curves are constructed for each material:

$$\theta_f = (t_f - t_0)/(t_{cr} - t_0) \leftrightarrow Z_v = \pi \cdot D \cdot n \cdot h \cdot c_l/\lambda_l$$

Development of the theoretical curves "$\theta_f \leftrightarrow Z_v$" is follows. For given values t_f and the chosen evaporating temperature of refrigerant t_0, the values of θ_f, θ_{in}, $\hat{r}_v$, $\hat{c}_{pl}$, $\hat{\lambda}_{pl}$ are calculated. The value $\pi \cdot D \cdot p/h^*$ depends on design of the apparatus: the size of the drum-evaporator (D); location of thermal measuring instruments, the technical ability to implement the continuous removing of the frozen product from the surface of the drum-evaporator (p). Substituting the calculated values of θ_f, θ_{in}, $\hat{r}_v$, $\hat{c}_{pl}$, $\hat{\lambda}_{pl}$ and ξ in the formula (3.67), we find the value of Z_v that appropriate to θ_f in the frame of the proposed PMM. The pair of numbers Z_v and θ_f defines one point on the coordinate plane Z_v - θ_f.

According to the experimental values of Z_v, one can determine the theoretically predicted final temperature on the product layer surface of the θ_{ft}. To do this, you must restore the perpendicular to the axis OZ_v (x-axis) from the point corresponding to the experimental value of Z_v, to the intersection with the theoretical curve for a given evaporating temperature of the refrigerant. The ordinate of the intersection point of the perpendicular to the curve equals the theoretical calculated final temperature $\theta_{ftheory}$. Then, the quantity that characterizes the divergence of the experimental results with the calculated data is defined: $(t_{ftheory} - t_{fexp})/t_{fexp}$.

The discrepancy between the experimental and calculated values in the range of admissible values of the similarity criteria and dimensionless conversion factors does not exceed 8%, which indicates the suitability of the proposed formula (3.66) to calculate the rational structures of DF.

Using this formula allows determining, according to the data of structural dimensions of the machine, its performance achievable for a given operating temperature, or vice versa, based on the required performance, to obtain the necessary relation between the structural characteristics of the device (see. chapter 4.1).

In addition, the achieved accuracy of the proposed generalized method of calculation eliminates the need for replacing it with partial criteria equations for each product separately.

It should be recognized that for this approach, although it achieves the goal, some characteristics of pragmatism are displayed. However, it should be emphasized that the holding of a numerical experiment based on the theory of active planning, provides an opportunity to represent the extent of "zone" in the form of a product of one-parameter functions of the selected factors. Thus there is confirmed the possibility of describing the processes of external heat transfer occurring in the drum freezers, based on the general physical concepts without the use of any empirical coefficients.

3.4 Calculation of the apparatus for freezing of pastor and mince-shaped products

As already mentioned (see Chapter 1), a large number of companies is engaged in the development of constructions of DF. However, information about them, given in the literature, as a rule, is promotional in nature, is not accompanied by scientific and technical arguments and only states the fact of the most rational organization of the thin layer freezing: liquid, paste- and stuffing-shaped, or small-piece products on a cooled cylindrical surface.

One of the leading companies in this field - the company "Atlas" - offers [69] the graph for determining the performance produced by its various types of drum units (Figure 4.1.) The graph is built for the initial (10°C) and the final (minus 20°C) temperatures of the product. The maximum thickness of the product in millimeters is marked. Then the perpendicular is put to the intersection with the line corresponding to the temperature of the cooling medium. From the point of intersection the horizontal line is carried to the intersection with the horizontal line corresponding to the frozen mass of the product. This mass is a specific, per 1 m² of the drum surface. So, there is calculated the time requiring for freezing the product to minus 18°C. Then, dropping a perpendicular to the x-axis, they define the capacity of different devices. One major drawback of this method is the need of constructing diagrams for each product separately.

However, experts from the same company proposed an analytical formula for calculating the performance of devices of the company "Rota - Freeze" for freezing of liquid and paste-shaped products.

Duration of freezing of the product layer with a given thickness is determined by the formula

$$\tau = (1 + 0{,}008\,t_{in}) \cdot \rho \cdot (i_f - i_{in}) \cdot (h/(2 \cdot k) - h^2/8 \cdot \lambda)/(t_0 - t_{cr}) \qquad (3.68)$$

where t_0 – refrigerant/coolant temperature, °C; t_{in}, t_{cr} - initial and cryoscopy temperature of the product, respectively, °C; i_f -i_{in} - the difference between the enthalpy corresponding to the amount of heat withdrawn during cooling of the product from the initial to the final temperature, kcal/kg; h -the maximum thickness of the product, m; k - heat transfer coefficient, kcal/(m·h·°C); λ - the thermal conductivity of the frozen product, kcal/(m·h·°C).

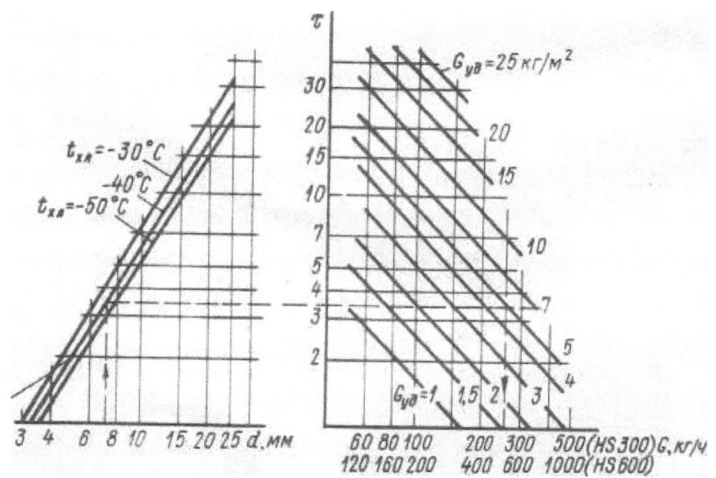

Fig. 4.1 Nomogram for determining the performance of drum machines of the company "Atlas" for types NBZOO and 115600 depending on the thickness of the product and the evaporating temperature of the refrigerant

Knowing the structural dimensions of devices the company "Rota-Freeze", we can calculate the achievable performance. However, this formula does not apply in the design of new constructions, rational search modes freezers, since it takes no account thermal characteristics of the material wall of the drum-evaporator, nonfrozen product as well as the inevitable reduction of the area of the working cooling surface with respect to the total surface area of the drum-evaporator.

In [66] there is suggested a method for calculating of DF for freezing pastor, minced-shaped products. It allows you to determine the design and operational characteristics of the specified performance device with the required precision.

For a predetermined capacity G, the adopted drum diameter D, the specific gravity of raw product ρ_l, the width of the mirror-surface of drum evaporator are known, b is defined by the formula

$$b = G/(3{,}600 \cdot \pi \cdot D \cdot n \cdot h \cdot \rho_l) \qquad (3.69)$$

Product n·h is numerically equal to the volume of food to be frozen and that transported through the area with a width of 1 m and a height h in 1 s. The value of n·h is determined from the relation

$$n \cdot h = Z_v \cdot \lambda_l/(\pi \cdot D \cdot c_l) \qquad (3.70)$$

In turn, Z_v is calculated by a formula connecting two parameters: 1. the dimensionless zone length of the product layer that is frozen and sub-cooled to the desired final (dimensionless) temperature θ_f; 2. the substantial main parameters of the investigated process

$$Z_v = \xi/[0.05 \cdot \theta_f^{-0.181} \cdot (\theta_{in} + 1{,}253) \cdot (\hat{r}_v + 1.617) \cdot \hat{c}_{pl}^{-0.05} \cdot \hat{\lambda}_{pl}^{-0.051}] - 25{,}64 \;(3.71)$$

Selection of the wall thickness of the evaporator is made only in according to the existing regulations of the Safety and/or Maritime Registers.

As an example, a calculation of the technical parameters of a drum freezer for freezing of fish pastor, minced-shaped products capacity of 12.0 tons/day is organized. Technical characteristics of the device are dictated primarily by the requirements of industrial fishing, and the lowest rate is determined as 100 kg /h. The desire for higher performance of one device, can lead to excessive increase in the diameter and width of the mirror of drum-evaporator. Such solution is offered by experts of Atlas company: the diameter of the drum-evaporator reaches 2 m, the width of the mirror is to 2 m. Requirements of designers restrict these dimensions. It is believed [22] that the maximum diameter of the drum-evaporator should be no more than 1.0-1.2 m, so take D=1 m. Performance of drum freezer is 12.0 metric ton/24h.

Currently fishing fleet is equipped mainly by refrigeration units that maintain the evaporating temperature of the refrigerant in freezers to minus 40°C. This temperature is taken as the operating temperature in the drum unit.

Assume that the product enters the apparatus at initial temperature that does not exceed $t_{in} = 20°C$, the $c_l = 3,57 \cdot 106$ J/(m³·K), $\lambda_l = 0,48$ W/(m·K), $t_{cr} = 1,7°C$, W = 81,7%, $\rho_l = 1,02 \cdot 103$ kg/m³ and the final temperature t_f product should be not higher than minus 18°C, respectively, $\omega = 81,5\%$.

As mentioned in Chapter 3.2, the possible variation range of the utilization coefficient of the cooling surface (p) is 0.51-0.75. We assume p = 0.6. It corresponds to 10% of the measurement error of thermophysical properties of the material to be frozen.

Using the above data, we have

$$\theta_{in} = (t_{in} - t_0)/(t_{cr} - t_0) = 1.566; \theta_f = (t_f - t_0)/(t_{cr} - t_0) = 0,574$$
$$\hat{r}_v = r_v \cdot W \cdot \omega/[c_l(t_{cr} - t_0)] = 1,626; \xi = \pi \cdot D \cdot p/h^* = 628 \ (3.72)$$
$$\hat{c}_{pl} = c_p/c_l = 1.098; \hat{\lambda}_{pl} = \lambda_p/\lambda_l = 34,588$$

Substituting (3.72) into (3.71), we find

$$Z_v = \pi \cdot D \cdot n \cdot h \cdot c_l \cdot \lambda_l = 1,470 \ (3.73)$$

By formula (3.73) it is easy to determine the relationship between the frequency of rotation of the drum-evaporator and the layer thickness of the product, ensuring the achievement of specified t_f at the respective predetermined parameters of the process: $n \cdot h = 62,92 \cdot 10^{-6}$ m/s.

From the formula (4.2), we find b = 0.8 m.

It should be noted that the apparatus for freezing pastor, minced-shaped fish products can also be used as an ice maker.

The above mentioned method of calculation can be used also for the calculation of drum apparatus used for purification of natural water precipitation, production of synthetic rubber latexes, the concentration of liquid products (including fruit and vegetable juices, coffee and tea extracts) by freezing.

REFERENCES

1. G.S. Altshuller. Creativity as an exact science. M.: Sovietskoe radio, 1979, 176 p. [in Russian]
2. M. Flemings. Solidification processes. Moscow: Mir, 1977, 423 p. [in Russian]
3. V.A. Zhuravlev, K.M. Kitaev. Thermo-physics of continuous formation of ingot. M .: Metallurgy, 1974, 216 p. [in Russian]
4. V.G. Boudina, M.A. Gromova. Thermal-physical characteristics of fish and products from it: Overview. Processing of fish and seafood. M .: TSNIITEIRH, 1977, vol. 1, 40 p. [In Russian]
5. E. Almasi, L. Erdelyi, T. Saray. Rapid freezing of food products. M.: Light and Food Industry, 1981. 408 p. [in Russian]
6. A.I. Veinik. Theory of solidification of casting. London: Mashgiz, McLaren and Sons Ltd, 1960, 435 p.
7. Rules for classification and construction of ships. Register USSR. L.: Transport, 1981. 960 p. [in Russian]
8. A.L. London, B.A. Seban. Rate of Jce Formation. Fransactions of the ASME, vol. 65, 1943, N 7, p. 771 - 778.
9. H.S. Carslaw, J.C. Jaeger. Conduction of heat in solids. Oxford: Oxford University Press, 1959, 487 p.
10. I. Stefan. Über die Theorie der Eisbildung, insbesondere über die Eisbildung im Polarmeere. Sitz her. Wien. Akad. Mat. Naturw., Bd 98, 1 la, 1889, p. 965 — 983.
11. L.I. Rubinstein. Stefan's problem, Riga: Zvaigzne, 1967, 457 p. [in Russian]
12. A.A. Gukhman. Introduction to the Theory of Similarity. NY: Academic Press, 1965, 296 p.
13. S.S. Kutateladze. Analysis of similarity in thermo-physics. Novosibirsk: Nauka, 1982, 280 p. [in Russian]
14. A.G. Tkachev, Convective heat transfer during solidification and melting of a homogeneous medium. Doctoral dissertation, L .: LTIHP, 1955.
15. M. Biot. Variational principles in heat transfer. Oxford: Oxford University Press, 1970, 209 p.
16. T. Goodman. Application of integral methods in nonlinear problems of unsteady heat transfer. In: Problems of heat. M.: Atomizdat, 1967, p. 41 - 96. [in Russian]
17. L.A. Kozdoba. Methods for solving nonlinear heat conduction problems. M.: Nauka, 1975, 228 p. [in Russian]

18. V.V. Salomatov. Methods for calculating the nonlinear processes of heat transfer. Tomsk: Tomsk state university, 1976, ch. 1, 245 p. [in Russian]

19. L.S. Leibenzon. Guide of oilfield mechanics. M.: GNTI, 1931, 335 p. [in Russian]

20. Analytical study of technological processes of cold meat. M.: TSNIITEIRH, 1970, 182 p. [in Russian]

21. V.B. Rzhevskaya. Investigation of the process of heat transfer in ice maker of continuous action to improve their performance. Doctoral dissertation, L.: LTIHP, 1976. [in Russian]

22. N.V. Fomin. Research and intensification of continuous ice machines. Doctoral dissertation, L.: LTIHP, 1974. [in Russian]

23. A .M. Makarov, V.A. Leonov, V.I. Dubovik, G.N. Shvedova. The problem of freezing liquid impingement on a flat wall. *Journal of Engineering Physics*, 1971, vol. XXI, no. 3, p. 537-546. [in Russian]

24. L.S. Leibenzon. On the question of curing the globe from its original molten state. Collected Works, vol. 3. Oilfield mechanics. M.: USSR Academy of Sciences USSR, 1955, 679 p. [in Russian]

25. A.A. Samarskii. Introduction to the theory of difference schemes. M.: Nauka, 1971, 552 p. [in Russian]

26. N.I. Nikitenko The study of heat and mass transfer by the method of nets. Kiev: Naukova Dumka, 1978, 213 p. [in Russian]

27. L.A. Kozdoba. Methods for solving problems of solidification. *Physics and Chemistry of Materials Processing*, 1973, no. 2, p. 41 - 59. [in Russian]

28. S.L. Kamenomostskaya. On the Stefan problem. *Mathematical collection*, 1961, vol. 53 (95), no. 4, p. 489-514. [in Russian]

29. G.V. Chizhov. Thermo-physical processes in refrigeration technology of food products. M.: Food Industry, 1979, 272 p. [in Russian]

30. V.A. Venikov. Simulation of energy systems. *Electricity*, 1971, no. 1, p. 5-13. [in Russian]

31. R. Plank. Modern refrigeration systems for the production of ice. *Refrigeration*, 1959, № 3, p. 10-15. [in Russian]

32. N. Shamsundar, E.M. Sparrow. Analysis of multidimensional conduction phase change via the enthalpy model. *J. Heat Transfer*, 97(3), 1975, pp. 333-340.

33. S.V. Zakrochinsky, M.D. Soekin. Guidance on the boiler supervision. M.: Metallurgy, 1963, 824 p. [in Russian]

34. B.F. Gromov, V.S. Petrishev. The solution of two-dimensional problems of hydrodynamics of a viscous incompressible fluid. *Proceedings of the All-Union Seminar on Computational Methods in Mechanics of a viscous liquid*. M.: Nauka, 1969, vol. 2, p. 74-87.

35. R. Siegel. Shape of Two-Dimensional Solidification Interface During Directional Solidification by Continuous Casting, *J. Heat Transfer* 100(1), 1978, p. 3-10.

36. T. Saitoh. Numerical method for multi-dimensional freezing problems in arbitrary domains, *J. Heat Transfer*, Trans. ASME. 100, 1978, p. 294-300.

37. A.N. Cherepanov. Assessing the impact of the axial component of the heat flux on solidification of the metal in continuous casting. *Applied Mathematics and Technical Physics*, 1977, no. 5, p. 102-107. [in Russian]

38. Z.M. Komladze. Investigation of heat transfer in continuous processes of freezing and freeze-drying of moist materials in thin monolithic layer. Doctoral dissertation, L.: LTIHP, 1974. [in Russian]

39. V.A. Berzin, V.N. Zhevlakov, Y.Y. Klyavin. Optimization of conditions of the ingot continuous solidification. Riga: Zinatne, 1977, 148 p. [in Russian]

40. V.P. Isachenko, V.A. Osipova, A.S Sukomel. Heat transfer. M.: Energy, 1975, 483 p. [in Russian]

41. Alyamovsky. Heat and mass exchange during cooling and storage of food. Doctoral dissertation, L.: LTIHP, 1974. [in Russian] I.G.

42. A.S. Ginsburg, M.A. Gromo, G.I. Krasouvskaya. Thermal characteristics of foods. M.: Food industry, 1980, 288 p. [in Russian]

43. N.A. Buchko. Algorithm for the numerical solution of two-dimensional Stefan problem by enthalpy method for three-layer explicit scheme. Refrigeration and Cryogenic Engineering and Technology. L: LTIHP, 1975, p. 142-154. [in Russian]

44. A.A. Samarskii. Introduction to the numerical methods. M.: Nauka, 1982, 277 p. [in Russian]

45. V.V. Skvortsov. Mathematical experiment in the theory of the development of oil fields. M.: Nauka, 1970, 224 p. [in Russian]

46. I.M. Vinogradov (Ed.). Encyclopedia of Mathematics. M.: Soviet Encyclopedia, 1977.

47. G.I. Marchuk. Methods of computational mathematics. M.: Nauka, 1980, 535 p. [in Russian]

48. A.I. Saltykov, G.L. Semashko. Programming for all. M.: Nauka, 1980, 157, pp. 158-160. [in Russian]

49. A.A. Samarskii, B.D. Moiseenko. Through calculation scheme for the multidimensional Stefan problem. *Computational Mathematics and Mathematical Physics*, 1955, vol. 5, p. 816- 827. [in Russian]

50. B.M. Budak, E.L. Solovyov, A.B. Uspensky. Difference method with smoothing coefficients for solving Stefan problem. *Computational Mathematics and Mathematical Physics*, 1965, vol. 5, no. 5, p. 828 - 840. [in Russian]

51. A.A. Gukhman. Introduction to the Theory of Similarity. NY: Academic Press, 1965, 296 p.

52. B. Menin. Calculation of error inherent physical-mathematical model due to finite number of recorded variables. *International Journal of Mathematical Models and Methods in Applied Sciences*, 7, 2013, 204-212. Available: http://goo.gl/j7ukzX.

53. B. Menin. Comparative error of the phenomena model. *International Referred Journal of Engineering and Science*, vol. 3, issue 11, November 2014, p. 68-76. Available: http://goo.gl/DwgYXY.

54. G. Greber, **S.** Erk, Grigull. Principles of Heat transfer. Translation from the German edited by Prof. A. A. Gukhman. M.: Foreign literature, 1958, 567 p. [In Russian]

55. S.K. Godunov, V.S. Ryaben'kii B.C. The theory of difference schemes. M.: Science, 1977, 440 p. [In Russian]

56. L.A. Dorfman. Numerical methods in gas dynamic turbo-machines. Leningrad: Energy Press, 1974, 272 p. [In Russian]

57. K. Hartman, E.V. Letsky, V. Schaefer. Planning experiment in the study of processes. M.: Mir, 1977, 552 p. [In Russian]

58. V.A. Venikov, A.M. Kuliyev. On the possible development of the theory of experimental design based on the theory of similarity. *Scheduling and automation of the experiment in research*/ed. G.K. Krug. M.: Soviet radio, 1974, p. 265 - 274. [in Russian]

59. V.V. Nalimov, N.A. Chernova. Statistical methods for planning of extreme experiments. M.: Nauka, 1965, 340 p. [In Russian]

60. A.L Lisenkov. Mathematical methods of planning multivariate biomedical experiments. M.: Medicine, 1979, 344 p. [In Russian]

61. E.N. Lvovsky. Statistical methods for constructing the empirical formulas. M.: Graduate School, 1988, 239 p. [In Russian]

62. V.L. Maksimov, V.D. Fedorov. Application of the methods of mathematical planning of the experiment in finding the optimal conditions for the cultivation of microorganisms. M.: MSU, 1969, 128 p. [in Russian]

63. L.A. Kozdoba, V.K. Miller. Numerical simulation of solidification of solutions. *Journal of Engineering Physics*, 1979, vol. XXXVII, no. 6, p. 750. [in Russian]

64. A.G. Badanov. Theoretical and experimental investigation of methods for constructing mathematical models of complex physical objects on the basis of experimental design. Doctoral dissertation, M.: Moscow Engineering Physics Institute, 1978.

65. A.V. Lykov, Y.A. Mikhailov. The theory of heat and mass transfer. M.-L.: Gosenergoizdat, 1963.536 p. [In Russian]

66. B.M. Menin, Study of heat transfer at freezing paste and minced products of sea fishing in the continuous devices and development of the method of their calculation. Doctoral dissertation, L.: LTIHP, 1981.

67. G.V. Maslova, E.L. Prudovskaya, I.R. Skomorovskaya. Manufacture of the cooked-frozen minced fish. M.: Food industry, 1978, 88 p. [in Russian]

68. V. Preobrazenskii. Thermal-technical measurements and instruments. M.: Energy, 1978, 703 p. [in Russian]

69. J. Brinch, P. Buus. Rotating drum freezer for continuous freezing of individual food. Proceedings of the XIV Int. Congress of cold. Vneshtorgizdat. M.: 1975, p. 276 -277. [in Russian]

70. A.A. Samarskii. Introduction to the theory of difference schemes. M.: Nauka, 1971, 552 p. [in Russian]

Printed by Books on Demand GmbH, Norderstedt / Germany